Managing Very Large IT Projects in Businesses and Organizations

Matthew Guah
Erasmus University Rotterdam, UK

INFORMATION SCIENCE REFERENCE
Hershey · New York

Director of Editorial Content: Kristin Klinger
Director of Production: Jennifer Neidig
Managing Editor: Jamie Snavely
Assistant Managing Editor: Carole Coulson
Managing Development Editor: Kristin M. Roth
Assistant Development Editor: Deborah Yahnke
Editorial Assistant: Rebecca Beistline
Typesetter: Lindsay Bergman
Cover Design: Lisa Tosheff
Printed at: Yurchak Printing Inc.

Published in the United States of America by
Information Science Reference (an imprint of IGI Global)
701 E. Chocolate Avenue, Suite 200
Hershey PA 17033
Tel: 717-533-8845
Fax: 717-533-8661
E-mail: cust@igi-global.com
Web site: http://www.igi-global.com/reference

and in the United Kingdom by
Information Science Reference (an imprint of IGI Global)
3 Henrietta Street
Covent Garden
London WC2E 8LU
Tel: 44 20 7240 0856
Fax: 44 20 7379 0609
Web site: http://www.eurospanbookstore.com

Library of Congress Cataloging-in-Publication Data

Guah, Matthew W., 1963-
Managing very large IT projects in businesses and organizations / by Matthew Guah.
p. cm.
Includes bibliographical references and index.
Summary: "This book offers authoritative research on the fundamental theory, practice, and implementation of very large successful IT projects in organizations"--Provided by publisher.
ISBN 978-1-59904-546-7 (hardcover) -- ISBN 978-1-59904-548-1 (ebook)
1. Information technology--Management. 2. Project management. I. Title.
HD30.2.G834 2009
004.068--dc22
2008030768

British Cataloguing in Publication Data
A Cataloguing in Publication record for this book is available from the British Library.

Table of Contents

Section IV: Case Studies

Foreword

'There are three routes to failure: gambling, sex and technology. Of these the first is the quickest, the second the most pleasurable and technology the most certain'
- Attributed to French President Pompidou

The writing has been on the wall for major IT or IT-enabled projects for a decade or more; range of factors makes them highly risky. Amongst these factors, three stand out. The first is size, the second is how far strong project management disciplines are applied, and the third is the degree of technology maturity (Willcocks, Petherbridge, and Olson, 2003). Low technology maturity refers to a situation where a new technology is being applied, or an existing technology is being applied in a new way, or where the knowledge and experience of the technology amongst the developers and implementers is low. In itself, low technology maturity can disable a major project. But when all three factors come together, as they do in some of the examples in this book, they form a heady risk cocktail indeed.

This is only the start of the book. Risk factors can also come from the history of IT in the organization. For example, a history of failure can make stakeholders sceptical about their ability to develop and manage the technology, let alone utilize information and communication technologies for business value. Failure, or at the very least, disappointment, then becomes a self-fulfilling prophecy—as can be seen in several major initiatives in the UK National Health Service from the late 1980s. Serious risk can also come from the external context supplier incapability, a business merger derailing the value and speed of development, for example. More obvious have been a series of factors relating to the internal context including lack of senior management support, lack of user buy-in, poor internal project management capabilities, lack of a supportive structure or culture of change, to name but a few. Then there are factors to do with the content and process of change. Therefore complex systems that have to be implemented across many business units with differing demands experience compound difficulties. Where a project touches many parts and users of an organization, the ability to manage the change process becomes an overriding critical success factor (Willcocks and Graeser, 2001).

There are many ways to deal with such issues, but, clearly, very detailed risk assessments followed by actions to mitigate the risks identified have to be at the forefront of management thinking. The question here is whether organizations are willing to broaden their risk assessments in order to take into account the new reality—the fact that large major IT projects these days are also business innovation projects and need to be managed as such. This brings into play more complex, intertwined risks, and the need to manage them through multiple actions and multiple stakeholders. Our own studies of major ERP and CRM projects find organizations and their IT suppliers experiencing serious difficulties with such a challenge, though we have also charted ways in which the more successful projects have been managed (Finnegan and Willcocks, 2007; Shanks, Seddon and Willcocks, 2003).

Clearly this book by Matthew Guah deals with a subject of great importance to contemporary economies and societies, to the point where, in the developed economies, one can say that no government agency or organization could avoid having to be knowledgeable about the issues this book deals with. Students of information systems will also find the book relevant to their studies, whether on undergraduate or postgraduate program. Having studied this area for some 20 years, and previous to that having been actively involved in many such projects for 10 years, I found the examples enlightening and the book highly topical. While there is still so much to learn about managing large-scale projects while the failure rate continues to change so little over so many years, and as such projects continue to be proposed, sponsored, and implemented, the need for a book such as this one will be recurring.

REFERENCES

Finnegan., D., & Willcocks, L. (2007). *Implementing CRM: From Technology To Knowledge.* Wiley, Chichester.

Shanks, G., Seddon, P., & Willcocks, L. (2003). *Second Wave ERP: Implementing For Effectiveness.* Cambridge University Press, Cambridge.

Willcocks, L., & Graeser, V. (2001). *Delivering IT and E-Business Value.* Butterworth Heinemann, Oxford

Willcocks, L., Petherbridge, P., & Olson, N. (2003) *Making IT Count: Strategy, Delivery and Infrastructure*. Butterworth Heinemann, Oxford.

Leslie Willcocks
Professor of Technology Work And Globalization
London School of Economics and Political Science

Leslie Willcocks is professor of technology work and globalisation at the London School of Economics and Political Science. His major research interests include outsourcing, IT management, large scale complex projects, such as CRM, ERP, organizational change and IT measurement. He is also engaged in looking at technology in globalization and the strategic use of IT, IT leadership, IT enabled organizational change as well as business process outsourcing and offshoring, organizational behaviour, social theory and philosophy for information systems, and public sector IT policy.

Preface

Why does it seem like every time I pick up a magazine or journal today, there is a leading story on the subject of very large IT projects (VLITP)? The subject of governments and multinationals trying, often miserably, to implement VLITPs has grown in both popularity and application. The academic community has also seemingly jumped onto this bandwagon with journals and conference papers (reporting justifications for VLITPs, quality management in VLITPs implementation, human resources management during VLITPs implementation, social and economical issues within VLITPs, amongst others). Regardless of what these papers and the popular press say about VLITPs, the final results appear to have similar emphasis. This book takes a cross-disciplinary and cross-functional approach towards VLITP implementation, enabling it to address the needs of the host organization. This includes aligning customer needs with business objectives to eliminate waste and increase costs, and to reduce cycle time, thereby driving the improvement objectives of VLITPs towards productivity that provides an opportunity to achieve the business goals—of both profitable revenue grown and market share expansion—for the host organization.

That VLITPs are beginning to play an increasingly central role in employment affairs; this is broadly acknowledged. Academicians in the 21st century are confronted with an entirely new challenge—to deliver a technologically-based alteration in cultural life, that is so rapid and all encompassing. Disappearing is the luxury of looking backward to understand the present, or even surveying the current terrain for a significant pattern. So variegated, chaotic, and rapidly proliferating are the changes in business culture that working practices can be little encompassed through rationally coherent analysis. The many attempts by VLITPs to change the way our world works, demonstrate that any analysis that is not already extended into the future is threatened with obsolescence at its moment of crystallization.

This book is different from the previous writings on the subject of project management primarily because it concentrates only on VLITPs. Moreover, it takes a more systemic or holistic organizational view by merging the contributions of all VLITP management issues into a single, unified approach to deliver breakthrough business process change as a result of implementing VLITPs.

In such spirit, the book offers several socio-technological arguments by both framing a single form of technological change as a result of implement VLITPs, and considering its implications for business practices in general and the management ethos in particular. It also proposes that by virtue of a confluence of VLITP emerging within this period of Internet age, businesses are immersed in a cacophony of competing claims for the real and good value of such technologies. In effect, there is a profound relativization of reality. With these conjectures in place, the book partly contributes to the concept of knowledge as a relational by-product, and to examine the implications of this view for the VLITPs as a pedagogical project in today's business environment. The author has argued that a transformation in the conception of knowledge heralds a radical reorientation to the strategic management of business processes through VLITPs. The book strongly suggests that this reorientation is already manifested, and it is anticipated that the present account may sharpen deliberation on the future directions of organizations efforts to manage VLITPs more successfully. In effect, this may be viewed as a transitional analysis, to be variously appropriated or refuted as VLITPs shape the future of innovation in different types of organizations and industries.

IMPACT

Traditionally, project managers have been allocated a project and their role has been to deliver on time, the quality standards and within budget (Youker, 1989). The host organization only recognizes the project outcome once the project is delivered—often leaving a gap between expectation and the final product. As the total budget of VLITP increases to uncontrollable amounts, project management's role is now changing and the total impact on the business needs to be addressed more effectively. The true role of a VLITP manager today may include:

- Acting as an orchestral conductor (which often requires demonstrating to the host organization an understanding of the VLITP's short, medium, and long-term objectives.
- Translating VLITP objectives into a something which the project team can satisfactorily address.

- Receiving from the many sub-projects a specification of the work to be undertaken, with the intention of selling its value to the host organization.
- Ensuring the host organization does not believe the VLITP can solve all their business problems.

Today, not only do most VLITP managers submit to the harsh logic of technical excellence for the final outcome, but also that the strategies for managing VLITP can be fundamentally redesigned to better serve the host organizations. That means a significant proportion of expectations, when evaluating the overall impact, should depend not primarily on technical innovations (Argarwal and Prasad, 1998) but also on concerns for fundamental issues like social philosophy (Wideman, 1998), the neutrality of technology (Mochal, 2003), and the related theory of technological determinism (Patzak, 1990).

If a VLITP delivers new technologies into an organization that is neutral, then its immense and often disturbing social and environmental impacts are accidental side effects of progress in the organization's business activities. This part of VLITP management functions deals with the current debate that polarizes around the question of whether the side effect of IS outweighs its benefits (Fish, 2002; Hagel, and Brown, 2001; Weill, 2004; Lacity, and Willcocks, 2006). Thus advocating the need to further progress in VLITP management strategies, claiming "strategic improvement" as an ally while the adversaries defend "business value" against emerging technologies and business process modernization as an environment wherein which the VLITP manager must strive to make a significant contribution. Some readers may consider impact factor—as explained in this section—to be a form of struggle for and against VLITPs that sometime do not seem to justify their apparent "colossal" budgets.

Critical Theory of IT rejects this observation and argues that the real issue is not the degree of improvements in IT *per se* but the variety of possible VLITP management strategies and the volume of IT projects among which businesses today must choose (Probert, 2002; Ray, 2001). IT is no more neutral than medieval cathedrals, but it embodies the values of a particular civilization and especially of its customary users, which rest their claims to hegemony on technical mastery. This book articulates and judges business values critique of VLITP. By so doing, the readers can grasp the outlines of another possible explanation for project management theories based on values to the host organization. This requires a different sort of thinking from the dominant technological rationality given to previous books on IT project management; a critical rationality capable of reflecting on the larger context of IT.

Throughout this book, it is shown that despite the support often voiced for staff participation in decision making when implementing VLITP, research provides no clear evidence that it results in higher productivity than management decision without participation. As a result of participation in decision-making, more demands are made on management than the other forms of employee involvement used within VLITP (Rodrigues and Hickson, 1995).

Several sections of this book analyze the roles of management teams for VLITP. The author shows how they need to take a number of actions to improve the decision-making during project implementations. This process starts with clearly defined objectives for all sub-projects, setting boundaries for each project manager and helps to lessen the dangers of outdated management beliefs having an influence on decisions. In addition to this, all managers on a VLITP should recognize and be aware of the limitations that exist which could affect the freedom of decisions. These limitations often stem from the host organization's political and social background, competition and economic scarcities, or even subordinates' attitudes to a particular project manager (Chang, Jackson and Grover, 2002; Dyer and Singh, 1998; Fiedler, Grover and Teng, 1995). Probably the most important concern about decision making, as far as VLITP is concerned, is the analysis of decision making in project stages to ensure that the project management process has formulated the reasons for taking a particular decision when problems occur during project implementation. This requires not just analyzing the nature of the problem but also examining alternative solutions and their possible consequences on the overall project's objectives. The author addresses these issues from several different angles in the chapters that follow.

This book should appeal to both academics and practitioners interested in the effective management of VLITPs wherein businesses. It is embedded in nested, interdependent networks of technical projects details as well as project management strategies that seem to be causing more obstacles for the success of VLITP. To adequately deliver its value, the author has given a balance and breadth of coverage to the necessary organizational literature. As a book for academics with groundbreaking work, the theoretical efforts of this author may invoke rather strong reactions—both positive and negative—from project management and organizational scholars.

CHAPTERS REVIEW

The book consists of 15 chapters, divided in three sections, visually represented in Figure P.1. Section I is made up of the first four chapters which frames and summarizes the book's contributions (see Figure P.1). Chapter I sets out the structure of the book by clearly defining the scope of a VLITP. It identifies vital characteristics that distinguish a VLITP from other types of IT projects. Chapter II defines a number of IS theories and situates them in relation to other approaches to VLITP management before giving details for various sources of project management mean-

Figure P.1. A visual representation of the book structure

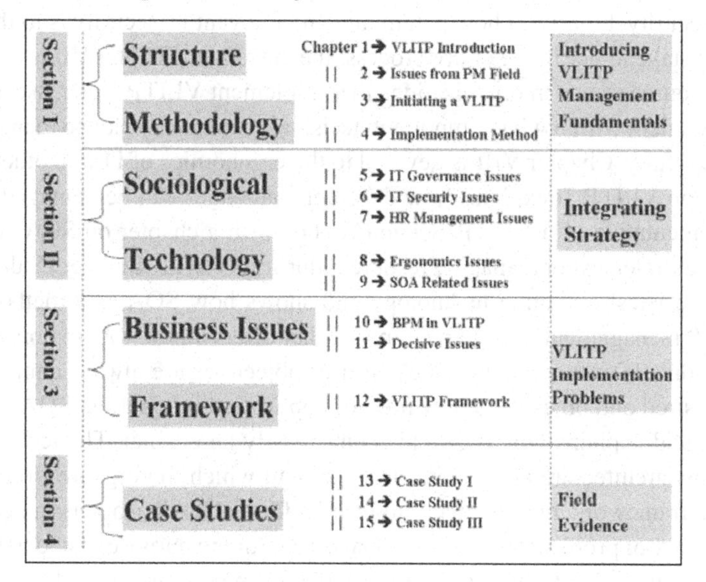

ings. The author argues here, that for all its insights in the 20th century, IS lacks a plausible strategy to change the way VLITP is managed. The historical experience of organizational improvement shows that businesses are not the primary agents of radical technological transformation, as Internet strategists believed (Guah and Currie, 2006). In Chapter III, the author establishes the value and reasoning behind the recent surge in implementing VLITP, while Chapter IV reviews various project management methodologies putting value to their significances for VLITP.

Section II consists of five chapters wherein various types of technologies involved in the implementation of VLITP and how the people that actually implement VLITP interact. These seemingly opposing factors of VLITP implementation are brought together in the discussion using management frameworks and principles that are formally developed and compared with conventional business management explanations for VLITP relative success or failure (see Figure P.1). Section II advances the on-going debate comparing relationship of human initiative to technical systems, both to general IT management and specifically to VLITP management situations. Since modernization of businesses is increasingly organized around VLITP, this relationship has become central to the exercise of management strategies.

Chapter V deals with recent improvements in IT Governance strategy, which can be used as strategic tools during VLITP implementation in the 21st century. It also investigates various IT control standards and provides evidence for their use in a VLITP situation. Chapter VI reminds the reader of the most risky aspect of any VLITP: security. It narrates how paramount and relevant IT security is to the most important stakeholder in the entire process: the customer. Chapter VII is about the human resource issues in organizations that implement VLITP. Such issues affect the environment within which important decisions have to be made to manage VL-ITP successfully. Chapter VIII is devoted to the ergonomics of IT and other social pressures on VLITP. It explains how different challenges can be managed during the implementation of a VLITP. Section II ends with a chapter on SOA and suggests some strategies for managing resources during a VLITP. Chapter IX describes SOA as the latest emerging technology and shows how SOA as a methodology for VLITP management is actually a means that offers a new way to construct an organization's IT infrastructure. SOA ensures objectives are always connected to VLITP desired outcomes. As a new form of application integration, SOA does not integrate; and explains how the architecture is only an enabler. The technologies used in this architecture are the enablers, without which the organization is only left with a framework, but no integration model for its IT. All chapters in Section II give a number of problems in VLITP management for the purpose of understanding how each of these relates to organizational actions in relation to VLITP.

Section III provides an overview of critical business issues involved in VLITP management. This section begins with Chapter X explaining the process stages of business and issues surrounding change management during the VLITP implementation. Chapter XI combines the most striking problems that can easily lead to the dismantling of a VLITP team and even abandonment. These are the issues of outsourcing and escalation during a VLITP implementation process.

This book considers the larger cultural context of technological change through the implementation of VLITP. Too often IT and culture are rectified and opposed to each other in arguments about the "trade-offs" between efficiency and substantive goals such as speed of implementation or participation and job security for existing staff or even compatibility to existing systems (Gerowitz et al, 1996; Mark, 1991; Williams, Dobson and Walters, 1990). A better understanding of the relationship between VLITP objectives and the overall goals of the host organization dissolves these apparent contradictions in chapter formulation of a VLITP management framework in Chapter XII.

Section IV provides an overview of VLITP management with evidence obtained from three different VLITPs, relating them to the theories in this book (see Figure P.1). All three chapters (XIII through XV) offer a deeper analysis of the histories of the illustrated project. Each demonstrates how VLITP management practices and procedures can change in response to different kinds of challenges. Chapter XIII focuses on the strategic role of VLITP in giving values to healthcare sectors in the United Kingdom. The chapter gives the reader detailed processes underlying the emergence and institutionalization of the stakeholder view. Chapter XIV focuses on the airline industry and similarly makes sense of the corporate strategies depicted in earlier chapters in light of the theoretical lens developed in Section I. Chapter XV highlights the descriptive and normative implications of this framework (see Chapter XII) to create value in VLITP implementation. It does this by detailing the implementation of OV-Chipcard in the Dutch public transport systems, where multiple companies collaborated in the implementation of a VLITP; forcing them to share costs and income from the final deliverable.

The book has been designed for use in a modular fashion. Students can refer to their syllabus and match certain chapters to the sequence presented in their specific syllabus. Practitioners, on the other hand, can also read it from start to finish and begin to experience the dynamics of the real world of management of VLITPs. Embedded in every chapter is a set of reflective questions to help readers link the theory with their own experience as well as examples and project situations, which runs throughout the book. Lecturers may find it useful for illustrating the key learning objectives in their syllabus.

The author's main objectives with this book are to update and extend the project management theory of IT, especially bringing VLITP management thoughts in line with current practices, and linking the theory to existing views of value creation and competitive advantage from new technology. To document current "best practices" in VLITP management, and how they were develop, the author has compiled an extensive case history on the largest non-military IT project by all conceivable standards:

- The National Health Services (provider of healthcare and services in the United Kingdom) national program for information technology currently referred to as the larges civil IT project, at the costs of US$ 10 billion (see Chapter XIII).

In making sense of this case, the author finds it helpful to draw on a wide range of organizational literatures (from IT governance to strategic business value). Particularly useful and central to this book, however, is a novel extension of recent work on VLITP implementation strategies with regards to service providers (Mark, 1991; Lacity and Willcocks, 2006; Williams, Dobson and Walters, 1990). The author highlights the instrumental and economic value of VLITP management in generating what could be business value from IT investment. Supported by the evidence from three cases, the author then details how and why business value from VLITP over the past two decades is reciprocally related to other forms of value being created by the host organizations for competitive advantage—such as the unique resource capabilities and industry-positioning advantages typically prescribed in the strategy literature.

REFERENCES

Argarwal, R., & Prasad, J. (1998). A Conceptual and Operational Definition of Personal Innovativeness in the Domain of Information Technology. *Information Systems Research,* 9(2), 204-215.

Chang, K., Jackson, J., & Grover, V. (2002). E-Commerce and corporate strategy: An executive perspective. *Journal of Information and Management*, 1-13.

Dyer, J.H., & Singh, H. (1998). The relational view: Cooperative Strategy and sources of interorganizational competitive advantage. *Academy of Management Review*, 23(4), 660-679.

Fiedler, K., Grover, V., & Teng, J. (1995). An Empirical Study of Information Technology Enabled Business Process Redesign and Corporate Competitive Strategy. *European Journal of IS*, *4*, 17-30.

Fish, E. (2002). An Improved Project Lifecycle Model, Pandora Consulting, http://www.maxwideman.com/guests/plc/intro.htm (Guest Department), 2002.

Gerowitz, M., Lemieux-Charles, L., Heginbothan, C., & Johnson, B. (1996). Top management culture and performance in Canadian, UK and US hospitals. *Health Services Management Research*, *6*, 69-78.

Guah, M. W., & Currie, W. L. (2006). Internet Strategy: The Road to Web Services Solution. Hershey, PA: Idea Group Publishing.

Lacity, M. C., & Willcocks, L. P. (2006). Transforming back offices through outsourcing: Approaches and lessons. In Leslie P. Willcocks and Mary C. Lacity *Global Sourcing of Business and IT Services*, Palgrave Macmillan, pp.97-113.

Mark, A. (1991). Changing Cultures—determining domains in the NHS. *Health Services Management Research*, *4*(3), 193-205.

Mochal, T., & Mochal, J. (2003). *Lessons in Project Management*, Apress, CA, p. 8.

Patzak, G. (1990). Project Management Paradigm: A System Oriented Model of Project Planning. *In Dimensions of Project Management*, edited by H. Reschke and H. Schelle, Springer-Verlag, Berlin, pp.26-27.

Probert, S. (2002). Combining Critical Theory with Empirical Studies in IS Research using Adorno's Critical Modelling Approach. *Proceedings of European Conference on Research Methods in Business and Management*, MCIL, Reading, pp. 309-316.

Ray, L. (2001). Pragmatism and Critical Theory. *European Journal of Social Theory, 7*(3), 307-321.

Rodrigues, S. B., & Hickson, D. J. (1995). Success in Decision Making: Different Organizations, Differing Reasons for Success. *Journal of Management Studies, 32*(5), 249-678.

Weill, P. (2004). Don't just lead, govern: How top performing firms govern IT. *MIS Quarterly Executive, 3*(1), 1–17.

Wideman, R. M. (1998). Dominant Personality Traits Suited to Running Projects Successfully (And What Type are You?). Project Management Institute, Annual Seminar/Symposium "Tides of Change", Long Beach, CA, USA.

Williams, A., Dobson, P., & Walters, M. (1990). *Changing Culture* (2nd ed.) London: IPM.

Youker, R. (1989, February). Managing the project cycle for time, cost and quality: Lessons from World Bank experience, *Keynote paper, INTERNET 88, Glasgow, 1988, 7*(1), 54.31

Acknowledgment

There many individuals who have contributed their expertise to improve the first draft of this book, and I am especially grateful to them. My first and foremost debt however is to Professor Leslie Willcocks who kindly agreed to discuss this book, and write the forward. I was extremely privileged to have had him as my internal examiner at Warwick Business School presently most fortunate to count him as one of most prestigious friends. Only those individuals fortunate enough to have been exposed to his special analytical discussions can appreciate the impact of his contribution to academic thinking.

Numerous other individuals have made contributions to this work. I am grateful to Michael Livesey (Royal Borough of Kingston-Upon-Thames in London) who generously gave his time during the incubation stage of this text and encouraged me to proceed with this project. Michael also provided me with moral support, discussed the results from reviewers and encouraged me to proceed with this project to completion. Students and colleagues at Erasmus School of Economics also assisted me with the verification of data and discussion of issues in this book, especially Professor H. Jeffrey Bouwer.

I am also very appreciative to all three anonymous reviewers, appointed by IGI Global to ensure my thoughts were appropriately aligned. Their painstaking reviews provided me with valuable suggestions which were incorporated in the text. To my friends at IGI Global I say a 'big thank you' for keeping the pressure on and moving things around to facilitate this book in time for a number of people to benefit. They include: Kristin Klinger, Julia Mosemann, Jan Travers, among others.

No work of this magnitude can be completed without the enormous sacrifices by the writer's family, and this is certainly true in my case. I am fortunate that my family fully understood the demands that the writing of this manuscript made on my leisure time—and that includes compassionate Eve, charming Michael and energetic Matthew.

Of course, the work on a text is never truly complete. There is always room for polishing and revision. I welcome comments from more knowledgeable and experienced individuals in the area of VLITP management who can help me to improve this text in subsequent editions.

Section I
Structure and Methodologies

Chapter I
Introduction to Very Large IT Projects

ABSTRACT

This chapter classifies the purpose of project management in IT projects as a means of introducing the topics covered in the book and demonstrates how a successful project manager must simultaneously manage these four basic elements of a very large IT project (resources, time, money, and scope). It also explains the impact of very large IT projects on business and the wider society today.

INTRODUCTION

Evidence that the size of an IT project is used to determine the extent to which project management practices are formally applied does not only come from the fact that sizing the project is a 'best-feel' technique among practitioners but also that it can be a scientifically derived factor. The size of an IT project guides the project manager through the application of project management practices helpful to that particular project (Reiss, 2007). Thus, embarking on a very large IT project (VLITP) requires that the host organization be aware of its "a priori" chances of success. Statistics of VLITPs failure rate provide a good measure of those chances (Cross, 2005). They are not shown to demoralize executives and to deter them from undertaking VLITP. This book is written to make project managers ponder on how to approach this endeavor so as to maximize their chances of success.

A successful project manager must simultaneously manage the four basic elements of a project: resources, time, money, and most importantly, scope (Archibald, 2003; Brown and Jones, 1998; Mochal and Mochal, 2003). The following elements of VLITP are interrelated and must be managed effectively for success:

- Resources (people, equipment, and material)
- Time (task durations, dependencies, and critical path)
- Money (costs, contingencies, profit)
- Scope (project size, goals, and requirements)

General IT project management literature emphasize the need to manage and balance the first three elements above, but the fourth element (scope) is most important for VLITP—the primary focus of this book (Kerzner, 1989; Patel and Morris, 1999; Webster, 1993; Youker, 1989). All VLITPs are expected to accomplish objectives based on project scope, restricted by the budget—of time and money. It is absolutely imperative that any change to VLITP scope has a matching change in budget, time and resources. Usually, scope changes occur in the form of "scope creep". Scope creep is the piling up of small changes that in a normal IT projects are manageable, but very significant for VLITPs. Within different stages of a VLITP minor changes can become a major addition without the equivalent adjustment in the overall project budget. Such situation can be best handled in a VLITP by ensuring any requested change—regardless how small—is accompanied by approval for a change in budget, schedule or both. VLITP cannot effectively manage the resources, time and money unless the project scope is actively managed (Bergeron and Bégin, 1989). When the project scope has been clearly identified and associated to the timeline and budget, the management of the project resources can begin. These include the people, equipment, and material needed to complete the project.

CLASSIFICATION OF IT PROJECTS

Project management practices help ensure that projects can be completed in a structured fashion – on time, on budget and producing expected results. Table 1 helps us to understand why one size may not fit all IT projects justifying the need for all IT projects to have a minimum level of project management strategy to ensure its success. The author refers to this classification (see Table 1.1) throughout the book to remind readers that project management process should not overtake the IT project and reiterates that applying the project management practices must consider differences in project size. When applied literally, Table 1.1 provides guidelines on possible roles of various participants in an IT project, dependent on the project size—obviously taking into these are only guidelines which can best serve as initial determining factors—in addition prior knowledge and experience of members in the project team.

Table 1.1 also shows how primary factors and impact factors must be balanced to determine project size. It has been argued that other factors (like duration, cost and project team size) cannot be the sole factors of project sizing (Martin et al,

Table 1.1. Characteristics and determinants of IT project size

Recommended Project Roles by Project Size	**Small**	**Medium**	**Large**	**Very Large**
PM level of involvement in all project activities	80 - 100%	40 - 80%	20 - 40%	> 20%
Customer liaison Officer	N/A	Optional	Yes	Yes
Customer Interest manager	N/A	Optional	Recommended	Required
Public Consultation	N/A	No	Optional	Recommended
No on PM team	<5 people	5- 50 people	50-100 people	> 100 people
Project Duration	<6 months	6 - 12 months	12-36 months	> 36 months
Total project cost	<$ 75,000	$75k - $500k	$500K-$5m	> $5m
Strategic value	Low impact	Minor	Major	Enormous
Impacts business units	Possibly none	1 - 5	5 - 10	>10
Impacts on end-users	Probably none	> 50	50 - 500	> 500
Stakeholder/management support	Optional	Low level	Moderate	Strong
Problem definition and solution achievement (involves interrelationship to existing systems, location of the implementation sites)	Easy to define problem and achieve solution	Somewhat difficult to define and achieve	Very difficult to define and achieve	Multiplicity seems nearly impossible to define and achieve
Dependencies on other projects	None	Minor links with low impact	Major links	Totally linked
Impact on financial revenue, expenses	N/A	Minor	Major	Significant
Familiarity of team with project's objective and solution (e.g., consider whether project team has ever used a required, new technology)	All team members know everything	Most team members are very familiar	Most somewhat familiar	Limited or no experience
Project Team very familiar with customer and their business	All team members know everything	Most team members are very familiar	Most somewhat familiar	Limited or no experience

continued on following page

Table 1.1. continued

Recommended Project Roles by Project Size	Small	Medium	Large	Very Large
Project Manager familiarity with project management practices (e.g., consider familiarity with containing scope creep, meeting firm deadlines)	PM knows everything	PM is very familiar	PM somewhat familiar	PM has limited or no experience
Project mission/vision defined before start date	Not necessary	Optional	Recommended	Required
Strategic alignment	Not necessary	Optional	Recommended	Required
Identify Stakeholders	Optional	Recommended	Required	Required
High level business requirement	Optional	Recommended	Required	Required
Feasibility of all sub-projects	Not necessary	Optional	Recommended	Required
Business case for each sub-project	Not necessary	Optional	Recommended	Required
Document approval for each sub-project	Not necessary	Optional	Recommended	Required
Project Scope defined	Recommended	Recommended	Required	Required
Define Milestone and allocate budgets	Optional	Recommended	Required	Required
Communication strategy	Direct with boss	Identify procedure	Document and implement	Document and implement
Quality Control Plan	Determine as you go	Identify procedure	Document and implement	Document and implement
Change management	Discuss with boss	Identify procedure	Document and implement	Document and implement
Risk management	Direct with boss	Identify procedure	Document and implement	Document and implement
Project kick-off meetings	Optional	Recommended	Required	Required
Gaining approval at various project stages	Not necessary	Recommended	Required	Required
Project rejection/ acceptance procedure	Recommended	Required	Required	Required
Intermediate Project Audit	Not necessary	Recommended	Required	Required
Transition period	>3 months	3-9 months	9-24 months	>24 months
Post-Project Audit	Not necessary	Recommended	Required	Required

2005). Yet nearly all project management researchers would agree on the fact that a typically lengthy, high-cost projects with sizeable project teams will be determined to be very large regardless of any specific impact factors. In contrast, a short-duration, low-cost project can still be technically complex and therefore attracting considerable interests from the senior management of the host organization. Such situation may explain why the application of small-project management practices may not be appropriate. In such case, a more appropriate strategy might be to adapt a project management approach that includes some more formal procedures than would ordinarily be suited for a small project. It is thus argued that primary factors and impact factors must be balanced to ensure the appropriate level of project management is applied to any IT projects.

The classification in Table 1.1 can lead to a usable VLITP management framework. The resulting framework would include all the activities needed to manage the entire IT project. Such activities would remain the same no matter the outcome (product or service) the VLITP delivers in the end. An important determinant is a suitable Product Development Life Cycle. This cycle includes activities to analyze, design, build and implement a specific product or service, which often varies depending on the type of product or service being developed.

RATIONALE FOR DISTINGUISHED PROJECT MANAGEMENT PRACTICES

This book incorporates two different levels of analysis of VLITPs—dealing with awareness and understanding. There are several differences between large and small IT projects. Being aware of these differences constitutes the first level of analysis. By explaining what gives rise to these differences the book suggests the second level of analysis, which involves understanding and appreciating the environmental and institutional peculiarities that uniquely affect VLITPs and influences the nature of the management functions. This level of analysis also involves explanations of the differences in how organizations deal with VLITPs.

Why do these differences exist? The analysis in this book focuses on how certain organizations operate within different stages of the life cycle of VLITPs. Table 1.1 illustrates that estimating the size of a project requires having a more easily constructible representative for the IT project in mind. Case study I (see Chapter XIII) demonstrates that knowing the size of the host organization, having good historical data on the relationship between the size of the previous attempts to solve the existing (business) problem and the eventual project, only then can it be possible to estimate the size of the current project.

A detailed look at the process of classifying a VLITP project shows a few shortcomings of the existing management processes:

- First, very little guidance is provided for a completely new type of project.
- Secondly, correct classification is necessary to avoid the temptation for a particular VLITP to either be over or under estimated.
- Finally, due to the lengthy process required for correctly categorizing a VLITP, it is usually an intricate task convincing top management that this will be project time/budget well spent.

CHAPTER SUMMARY

Managing VLITPs is proving to be increasingly difficulty. It has become so important to large businesses today because majority of VLITP in recent years have led either to complete failure or over expenditure—partly due to the need for strategic simplification. Several proprietary methodologies mentioned in this book put the finality of VLITP in its right perspective. VLITPs have unique characteristics that often turn them into serious jobs for the host organizations' future success. Defining the success or failure of a LITP is always tricky given its far-reaching implications into a commercial future of the host organization. The failure of implementation strategy should always be considered to avoid failure of the entire organization. Its effects will not wither away by cutting off a piece of that organization. Instead it is only a holistic review of the progress can ensure chances of success. The consequences of VLITP failure on the other hand can be astounding and have far reaching repercussions for the future of the IT industry.

This is an important book for helping the reader to understand many important questions currently coming out of IS research, typically addressing corporate move towards implementing VLITP, in direct relations to reforming different kinds of organizations. Reforming organizations with technology, has long been in search of a credible theoretical link to performance measurement (Currie and Guah, 2006). This book is an impressive step toward that goal. The author manages to usefully put together the challenges, social concerns of stakeholder management models with the economic focus of the resource-based and industry-structure views of competitive advantage during the implementation of VLITP. In doing so, this book puts a more humanistic face on the currently dominant economic explanations of competitive advantage that VLITP brings to the host organization and takes IT implementation and project management theories to new heights of mainstream applicability.

REFERENCES

Archibald, R. D. (2003). Managing High-Technology Programs and Projects, Third Edition, Wiley, p. 44.

Argarwal, R., & Prasad, J. (1998). A Conceptual and Operational Definition of Personal Innovativeness in the Domain of Information Technology. *Information Systems Research, 9*(2), 204-215.

Bergeron, F., & Bégin, C. (1989). The Use of Critical Success Factors in Evaluation of Information Systems: A Case Study. *Journal of Management Information Systems, 5*(4), Spring, 111-124.

Brown, A. D., & Jones, M. R. (1998). Doomed to Failure: Narratives of Inevitability and Conspiracy in a Failed IS Project, Organization Studies, 19(1), 73-88.

Chang, K., Jackson, J., & Grover, V. (2002). E-Commerce and corporate strategy: An executive perspective. *Journal of Information and Management*, 1-13.

Cross, M. (2005, October). Special Report: Public Sector IT Failures. *Prospect,* 48-52.

Currie, W. L., & Guah, M. W. (2006). Web Services in National Healthcare: The Impact of Public and Private Collaboration. *Information Systems Journal, 1*(2), 48-61.

Dyer, J. H., & Singh, H. (1998). The relational view: Cooperative Strategy and sources of interorganizational competitive advantage. *Academy of Management Review*, 23(4), 660-679.

Fiedler, K., Grover, V., & Teng, J. (1995). An Empirical Study of Information Technology Enabled Business Process Redesign and Corporate Competitive Strategy. *European Journal of IS*, *4*, 17-30.

Fish, E. (2002). An Improved Project Lifecycle Model, Pandora Consulting, http://www.maxwideman.com/guests/plc/intro.htm (Guest Department).

Gerowitz, M., Lemieux-Charles, L., Heginbothan, C., & Johnson, B. (1996). Top management culture and performance in Canadian, UK and US hospitals. *Health Services Management Research, 6*, 69-78.

Guah, M. W., & Currie, W. L. (2006). Internet Strategy: The Road to Web Services Solution. Hershey, PA: Idea Group Publishing.

Hagel III., J., & Brown, J. S. (2001). Your next IT Strategy. *Harvard Business Review,* 79(9), 105-113.

Kerzner, H. (1989). Project management: A Systems Approach to Planning, Scheduling, and Controlling, 3rd Edition. NY: Van Nostrand Reinhold. p. 84.

Lacity, M. C., & Willcocks, L. P. (2006). Transforming back offices through outsourcing: Approaches and lessons. n Leslie P. Willcocks and Mary C. Lacity *Global Sourcing of Business and IT Services*, Palgrave Macmillan, pp. 97-113.

Mark, A. (1991). Changing Cultures – determining domains in the NHS. *Health Services Management Research*, *4*(3), 193-205.

Martin, N. L., Pearson, J. M., & Furumo, K. A. (2005). IS Project Management: Size, Complexity, Practices and the Project Management Office. *38th Annual Hawaii International Conference on Systems Sciences*, Hawaii, USA.

Mochal, T., & Mochal, J. (2003). *Lessons in Project Management*, Apress, CA, p. 8.

Patel, M. B., & Morris, P. G. W. (1999). *Guide to the Project Management Body of Knowledge*. Center for Research in the Management of Projects, University of Manchester, UK, p-52.

Patzak, G. (1990). Project Management Paradigm: A System Oriented Model of Project Planning. In *Dimensions of Project Management*, edited by H. Reschke and H. Schelle, Berlin: Springer-Verlag, pp. 26-27.

Probert, S. (2002). Combining Critical Theory with Empirical Studies in IS Research using Adorno's Critical Modelling Approach. *Proceedings of European Conference on Research Methods in Business and Management. MCIL*. Reading, pp. 309-316.

Ray, L. (2001). Pragmatism and Critical Theory. *European Journal of Social Theory*, *7*(3), 307-321.

Reiss, G. (2007). Project Management Demystified: Today's tools and techniques, (2nd ed.), USA: Taylor and Francis Ltd.

Rodrigues, S. B., & Hickson, D. J. (1995). Success in Decision Making: Different Organizations, Differing Reasons for Success. *Journal of Management Studies*, 32(5), 249-678.

Webster, F. M. (1993). What Project Management Is All About? In *The Handbook of Project Management* edited by Paul C. Dinsmore, Amacom, NY, p. 8.

Weill, P. (2004). Don't just lead, govern: How top performing firms govern IT. *MIS Quarterly Executive*, 3(1), 1–17.

Wideman, R. M. (1998). Dominant Personality Traits Suited to Running Projects Successfully (And What Type are You?). Project Management Institute, Annual Seminar/Symposium "Tides of Change", Long Beach, CA, USA.

Williams, A., Dobson, P., & Walters, M. (1990). *Changing Culture* (2nd ed.) London: IPM.

Youker, R. (1989, february). Managing the project cycle for time, cost and quality: Lessons from World Bank experience. *Keynote paper, INTERNET 88, Glasgow, 1988, Volume 7*(1), 54.31.

Chapter II
The Field of Project Management

ABSTRACT

By examining the history of what was earlier considered project management, this chapter not only points out lessons from past practices but also justifies the selected definition of VLITP. It also explains the role of project management in a fast business environment. The author has demonstrated such importance by representing VLITP in the form of a major initiative that contains a series of relevant processes in the host organization.

INTRODUCTION

This chapter looks at the general definition of project management and what is required to manage VLITPs. It begins by explaining the role of project management before leading to specific details relating to VLITPs. Relating project management to our everyday lives the book gives some example of how experiences of none IT activities can contribute to the effective management of VLITP. It also shows how business environment demonstrates a particular view of an organization—primary objective is making profit—but relates this to other consideration that often seem very much subservient to this. Before concluding with a futuristic look at project management, the book provides an in-depth understanding of project, program and portfolio management and describes how this had been designed to satisfy the

Figure 2.1. Dealing with the Beast of VLITP

needs from the top executive to the junior practitioner of the business within which a VLITP is being implemented.

Having completed a success VLITP is not proportional to being a successful project manager. Reiss (2007) list several projects that terribly managed by either completing years late or costing twice as much as the worst estimate, and yet the project manager managed to get promoted out of sight. On the other hand certain projects see devoted and capable people fight against enormous odds but yet achieve completely the wrong objectives. This goes to emphasise that there are twin objectives in project management—completing the project successfully and you (the project manager) have a future, a career and hopefully an enormous salary. To achieve these goals every manager of a VLITP must learn to avoid problems. Managing VLITP is therefore about tackling new grounds, taking a group of people and trying to achieve some not so clear objectives as quick as possible and as efficient as you can.

Many VLITPs fail due to their complexity, size and duration. They introduce new challenges in the filed of project management, resulting to new problems. Due to the *nature of this beast* (see Figure 2.1) managing VLITP tend to be unpractised and unrehearsed. The process of managing a VLITP involves time and money—which often grow and grow, becoming more and more threatening, often becoming impossible to handle (see Figure 2.1). For this very reason, this book suggests that a methodological approach to VLITP could reduce the current very high rate of failures. VLITP today means different things to different people at different levels in the organization hierarchy. But for most, a VLITP has a large and direct impact on the following:

- Processes within a business—often these are primary and secondary processes which also includes people management
- The systems that run and support business processes
- Products and output that the system is delivering (core or resultant).

WHAT IS PROJECT MANAGEMENT?

This section clarifies the purpose of project management by first looking at the origin of project management. It then looks at the formulation of the objectives and policies of a business and the pursuit of all the necessary activities that help in to achieve such objectives and policies.

The earliest forms of what is considered as project management today can be found during the industrial revolution (Kaushik and Cooper, 2000). The roots of the industrial revolution began in the United Kingdom in the late eighteen-century, then the most powerful nation on earth—ruling empires and holding the political strength with all of the components necessary to sustain a revolution of such magnitude that it would change the face of the planet for all time. The UK had wealth, entrepreneurs, and financial markets to promote new automation, factories, and manufacturing and management techniques. After managing industrial projects started in the United Kingdom, the revolution quickly spread, first to France and Germany, then eventually to the United States of America.

This was a period of great innovation that lasted for a period of about 150 years. Few of the many advances in technology, which managed to survive the changes that have occurred around the globe since that period, are listed below:

- Cement was invented by an Englishman called John Smeaton in 1756.
- The steam engine was invented by James Watt in 1770.
- The spinning jenny used in textile manufacturing was invented by James Hargreaves in 1764.
- The railway locomotive was invented by George Stephenson in 1814.
- The Bessemer steel process was invented by Henry Bessemer in 1856.
- William Siemens invented the electric furnace in 1861.

All these started as little projects by a single individual but grew into large projects with global objectives mainly because they supported the outmost objective of any business endeavour—to maximize profit. That is because *nearly* every entrepreneur is urged to initiative project as a business activity to find an opportunity to fill a market need, or to create one, which brings financial rewards.

Three centuries later, project management is mostly concentrated on IT-projects with various definitions. This book uses, for its working definition, the Project Management Institute (PMI) chosen definition (see PMI website):

"A project is a temporary endeavor undertaken to accomplish a unique purpose. Project management is the application of knowledge, skills, tools and techniques to project activities in order to meet project requirements."

Almost any human activity that involves carrying out a non-repetitive task can be a project. We are all project-managers since we all practise project management in one form or the other. But there is a big difference between carrying out a very simple project involving one or two people and one involving a complex mix of people, organisations and tasks. This has been true for millennia, but large-scale projects like the building Noah Ark—according to the Holy Bible—often use rather simple control and resource techniques including brute force to 'motivate' the workforce.

The art of planning for the future has always been a human trait (Eyre and Pettinger, 999; Stewart, 1997). Early version of project management strategy captured on paper the essence of a project with a couple of simple elements including:

- A start date.
- An end date.
- A good description of the tasks that have to be carried out.
- An indication of when each task should be finished.
- An indication of idea of the resources—including people, machines, etc—that was needed during the course of the project.

When the plan of a project starts to involve different things happen at different times, often depending on each other, on resources required at different times and in different quantities. Working at different rates, the paper plan would start to cover a vast area and be unreadable. This was a problem faced the US Navy in the development of the Polaris missile system that consisted of so many aspects to the project to justify the invention of a new technique to cope with it: the PERT technique. As a result of improvising and later developments in project management procedures, this coincidence led to mathematical techniques that can be used to find the critical path through a series of planned tasks that interconnect during the life of a project.

While some may claim that the entire story of modern project management began from this time, the author considers such assumption to be rather unfair as project management is not only about planning but also about certain human attributes like leadership and motivation. Nevertheless, the important lesson from the project management history is the idea of complex plans could be analysed by one IT system to allow someone to control a project—the basis of much of the development in information systems. That idea has now been developed further that it allows for projects of any size and complexity not only to be planned but also modelled to answer the 'what if?' questions. Although the original impression given by IT projects tended to produce answers long after an event had taken place, the 21st century introduced many project planning and scheduling applications that can provide real time information. Thus, linking project management issues to risk analysis, time recording, costing, people evaluation, phrase estimation and other aspects of project control.

IT applications alone are not project management but only tools that enable project managers to do their jobs easily and efficiently. Good management of VLITP is a mixture of various components—of control, leadership, teamwork, resource management, etc—that goes into a successful project. VLITP managers are not only found in all industries but also expected to contribute in these industries. As a result the number of VLITPs has grown rapidly within an increasing list of industries. Business appreciation of the necessary skills for a well-managed VLITP has also been greatly improved. As more VLITPs started to emerge within big businesses and public organisations in the last decade, the management of VLITPs has become well established as both a professional career path and a way of controlling business processes and functions. This means opportunities in VLITP now exist not only in being a project manager, but also as part of the support team in a sub-project or programme office or even as a team leader for a sub-project. Many Internet sites currently list various qualifications that can be obtained by becoming associated with a number of professional associations. One reason for this rapid growth is the need to understand how to look after VLITP, which are often very complex, often in high tech areas, and not only critical to business success but also the need to efficiently manage scarce resources.

Managers—these are people—who carry out management, thus making it necessary for this book to include discussions of managers and their roles in VLITP. Existing literature contains examples of many people who are been given the title of managers today but do not quite have what it takes to be one (Brown, 2001; Daniels et al, 2002; Kanter, 2001; Kaushik and Cooper, 2000; Markus, 1994; Reddy, 2004; Stewart, 1997). These people could generously be given the description of supervisors, team leaders or simply be called administers. This book refers to these people as supervisors because they make decisions in accordance with rules that are laid down, with very little or no discretion being required or indeed, allowed. In cases where problems arise that cannot be resolved within the established rules, these people have to refer to their superiors. A manager is a person who has the authority to use his or her discretion in making decisions and the limits to this discretion indicate the manger's place on the management ladder within the organization. This goes to explain why there are several attributes that make up the qualities required for VLITP management. A number of the desirable characteristics listed by Eyre and Pettinger (1999) are: self-confidence, drive, initiative, decisiveness, willingness to accept responsibility, ability to delegate, integrity, judgement, adaptability, organizing ability, stamina, emotional maturity, human understanding, personality, being supportive of staff, and adequate educational standard.

There still exists the classic time, quality, and cost triangle. That's because most people still want their projects to be on time, meet quality objectives, and definitely cost far less than the agreed budget (Youker, 1989). However, if an organization has an unlimited budget and unlimited time, the job of a project manager becomes rather easy. Unfortunately for most organizations, time and money are critical and that results to an increase value in the job of a project manager today.

Risk is a key issue for every project manager (Currie, and Willcocks, 1998; Sagripanti et al, 2002; Slay, 2006; Smith, 2000; Vincent et al, 1998; Westney, 2001). How the various risks are handled can often define the final outcome of a project; considering it often determines the overall worth to all parties to the project and ultimate the success or failure of the project. In our effort to lift the lid on the real reasons why problems with VLITPs exists the book forces organizations to think outside the box—by considering the fundamental reasons why they persistently fail to achieve the success they seek from their information systems projects. VLITPs can often be extremely complex processes involving various teams from client organizations, IT service providers and sub-contractors. Making sure you don't fall foul of the law may not be uppermost in the project manager's mind. The law in most countries for Project Managers provides an easily understandable and practical guide to the laws of contract, liability, intellectual property and so on, entirely from the perspective of the project manager (Currie, 2003; Smith, 2000). This enables the project manager to approach projects forewarned and forearmed, primarily to avoid potential legal problems entirely.

Project Management–A Process

Whether project management is an art or a science isn't the main concentration of this book. Project management is a process used to accomplish organizational goals through the implementation of a project; that is, a process used to achieve what an organization wants to accomplish by undertaking a particular project. Project managers are the people to whom this management task is assigned, and it is generally thought that they achieve the desired goals through the key functions of:

- Planning.
- Organizing.
- Directing.
- Controlling.

The four key functions of project management are applied throughout an organization regardless of whether it is a business, charity, or a government department (Feldman, 2004). In a VLITP for a business, many different subprojects with various activities take place (i.e. buy merchandise to sell, sell the merchandise, prepare the merchandise for display, advertising and promotion, do the accounting work, hire and train new employees, etc). There might be one project manager for the entire activity, but there are other project managers at different levels who are more directly responsible for the people who perform all the other subprojects. At each level of VLITP implementation, the four key functions of planning, organizing, directing, and controlling are included. The emphasis changes with each different level of manager, as will be explained later.

Planning: Planning in any organization occurs in different ways and at all levels. A top-level project manager plans for different events than does a project manager who supervises a group of workers who are responsible for assembling modular homes on an assembly line. This project manager must be concerned with the overall operations of the full project, while the assembly line manager or supervisor is only responsible for the line that relates directly to that sub-project.

Planning could include setting organizational goals for the VLITPs—usually carried out by higher-level project managers on a VLITP. As a part of the planning process, the project manager then develops strategies for achieving the goals of the project desire for the host organization. In order to implement the strategies, resources will be needed and must be acquired. The planners must also then determine the standards or levels of quality that need to be met in implementing the VLITPs (Feldman, 2004). The three general categories of VLITP planning are: strategic planning, tactical planning, or contingency planning:

- Strategic planning is long-range planning that is normally completed by top-level managers in an organization.
- Tactical planning is short-range that is done for the benefit of lower-level managers, since it is the process of developing very detailed strategies about what needs to be done, who should do it, and how it should be done. To return to the previous example of assembling modular homes, as the home is nearing construction on the floor of the plant, plans must be made for the best way to move it through the plant so that each worker can complete assigned tasks in the most efficient manner. These plans can best be developed and implemented

by the line managers who oversee the production process rather than managers who sit in an office and plan for the overall operation of the company. The tactical plans fit into the strategic plans and are necessary to implement the strategic plans.

- Contingency planning allows for alternative courses of action when the primary plans that have been developed don't achieve the goals of the organization. In today's economic environment, plans may need to be changed very rapidly. Continuing with the example of building modular homes in the plant, what if the plant is using a nearby supplier for all the lumber used in the framing of the homes and the supplier has a major warehouse fire and loses its entire inventory of framing lumber. Contingency plans would make it possible for the modular home builder to continue construction by going to another supplier for the same lumber that it can no longer get from its former supplier.

Organizing: Organizing refers to the way the resources are allocated and VLITP resources, assigns tasks, and goes about accomplishing its goals. In the process of organizing, project managers arrange a framework that links all members of sub-projects, workers, tasks, and resources together so the organizational goals can be achieved (Angell and Smithson, 1991). Guah (2007) refers to this framework as organizational structure of IT projects. It is necessary to have an organizational chart explains the organizational structure of VLITP. The chat depicts the structure of the project organization showing positions in the project, usually beginning with the top-level manager (normally the person spearheading of the project) at the top of the chart. Other managers are shown below the project director.

It is important to note that the choice of structure is important for the type of environment in which the project is being implemented, its clientele, and the products or services it provides—all which influence the overall goals of the host organization.

Directing: Directing is the process—many would relate to managing VLITP—of supervising, or leading workers to accomplish the goals of the project team. In many VLITPs, directing involves making assignments, assisting workers to carry out assignments, interpreting organizational policies, and informing workers of how well they are performing. To effectively carry out this function, project managers must have leadership skills in order to get workers to perform effectively.

Some VLITP managers direct by empowering workers. This means that the manager doesn't stand like a taskmaster over the workers barking out orders and correcting mistakes. Empowered workers usually work in teams and are given the authority to make decisions about what plans will be carried out and how. Empowered workers have the support of project managers who will assist them to make sure the overall goals of the VLITP are being met. It is generally thought that workers who are involved with the decision-making process feel more of a sense of ownership in their work, take more pride in their work, and are better performers on the job.

By the very nature of directing, it should be obvious that the project manager must find a way to get workers to perform their jobs. There are many different ways project managers can do this in addition to empowerment, and there are many theories about the best way to get workers to perform effectively and efficiently during VLITP.

Controlling: The controlling function of managing VLITP involves the evaluation activities that project managers must perform. It is the process of determining if the VLITP's overall goals and objectives are being met. This process also includes correcting situations in which the goals and objectives have not being met. There are several activities that are a part of the controlling function (Kerzner, 1989).

Project managers must first set standards of performance for VLITP workers. These standards are levels of performance that should be met. Such standards must then be communicated to sub-project managers who are supervising workers at various levels, and then to the workers so they know what is expected of them as part of the VLITP.

After the standards have been set and communicated, it is the project manager's responsibility to monitor performance to see that the standards are being met. If the project manager watches the VLITP go through the various stages (see Figure 2.2) and notices that it is taking longer then planned, something must be done about it. The standards that have been set are not being met. A good VLITP manager would find it relatively easy to determine where the delays are occurring. Once the problems are analyzed and compared to expectations, the project manager must do something to correct the results. Normally, the project managers would take corrective action by working with the employees responsible to put things right. Considering there could be many reasons for the delays, several options should be investigated until the appropriate corrective action has been taken.

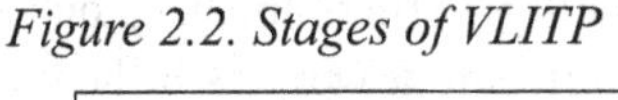

Figure 2.2. Stages of VLITP

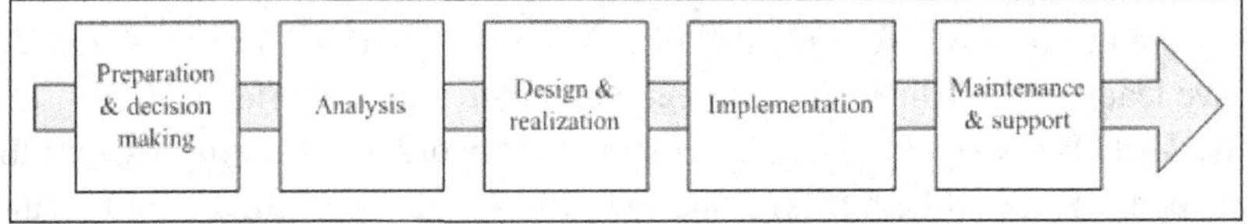

ASSESSMENT OF PREVIOUS PROJECT MANAGEMENT PRACTICES

Regardless of the original motivation for a particular business, the eventual growth put more emphasis on profit maximization. Such need to maximize project often leads to the pursuit of VLITP as a means of achieving the shareholders wish to see the maximum possible return on their investments—particularly true of large institutional investors. Business environment demonstrates a particular view of an organization which is that organizations exist to make a profit and that all other consideration are very much subservient to this. This is referred to as *economic view* (Eyre & Pettinger, 1999: p.8). The implication of such view is that businesses are implementing VLITPs and managing them appropriately to foster the fact that they are in businesses to mix the factors of their product and services in the proportions appropriate to its particular activities so as to make their products and services sellable at the maximum profit.

Previous project management, in regards to VLITPs, have taken in consideration the growing intervention of governments around the world in the operations of business activities. Varies types of laws and regulations—relating to employment procedure, location of offices, etc—not only increase the costs of managing VLITPs but also influence the means of implementing them. Nearly all VLITPs directly affect the future of competitions in one industry or another. The fact that certain industries (i.e. health, defence, and utility) are restricted, implementing VLITPs in them is considered to have unfair influences in the form of public or private monopolies (Currie & Guah, 2007). Eyre and Pettinger (1999) effectively describe this as economic view—which doesn't only realise efficiencies of VLITP management for profit maximization in the short-term but also demonstrates that success in profit maximization in the long-term can be optimised by successfully implementing VLITPs. The problem with this view is that very little allowances are made for changes in circumstances or emerging technologies or even political situation.

Managing VLITP requires specific attention to varies stakeholder who are very much interested in the formulation and achievements of specific business objectives. The most well known stakeholders are as follows:

- The owners or investors in the business and in the case of public sectors are the government and its departments and agencies.
- The directors and departmental managers who have been given the responsibilities for delivering the benefits of other stakeholder which occasionally conflict with their own interest.

- The suppliers who usually are dependent upon the continued well-being of the business for its own success.
- Similar to the suppliers are the interests of the distributors.
- Finally but usually the most important stakeholder is the customers who do not only demand continuous increase in quality of product and services but also includes the local and wider community with diverse opinion of the product and services.

The above represents the often diverse and conflicting interests of those with vested interest in the implementation of VLITPs by businesses around the world.

Various policies in business today are shaped out of a combination of objectives, individual drives and social demands (Brown and Jones, 1998; Galliers, 1998; Reddy, 2004). Whether a specific business policy will bring financial gain or merely a social benefit to the organization depends on the fact that the policy is based on economic objectives or social considerations. Reflecting on the views of the above stakeholders, the policies of business to implement VLITPs should take a broad view seeking to address various concerns as well as ensuring profit in the long-term. This means VLITP overall management must take on board the seriousness of successful implementation. This also includes the support of all directors of the organization including non-executive directors where they exist. Together they need to consider social and environmental aspects of implementing VLITPs.

Project management contributes to the way businesses tackle the complex and sometimes intimidating environment of VLITP implementation. Only if a business understands and can implement a really good project management strategy could it ensure a successful VLITP. Therefore project management gives copious graphic illustrations and bullet lists that should meet the needs of businesses searching for a quick understanding of the major issues involve during the implementation of various types of VLITPs.

Project managers over the years have used their many years of project-based experience to provide an in-depth understanding of projects, program and portfolio management, usually designed to satisfy the needs from small to large business executives—usually by emphasising the quality of project outcome. On the one hand, VLITP quality is the collective name given to a set of practical tools and methods that work well and make sense; yet, it is a concept that brings about change, sometimes rendering acceptable improvements that would be resisted if they were introduced in the name of cost cutting or greater productivity.

VLITP managers use various different approaches to quality management—mostly based on systems techniques appropriate to the needs of individual businesses—linked coherently rather than emphasizing any single approach. Project managers, over the years, have demonstrated great interests in improving organizational performance by managing projects' quality more effectively (Kerzner, 1989; Patel and Morris, 1999). They have not been confined to particular areas of commerce, industry or the public sector but a continual development and enhancement, which may differ somewhat in content and structure from the outline given below:

- *Approaches to quality* of project management over the last decade may differ but organizational implications of the total quality approach to VLITP management continues to include the application of systems thinking and regards for legislation.
- *Quality economics* involves cost models that are devised by obtaining and processing cost information. Such models use cost information to improve quality performance in an attempt to satisfy the host organization while reducing costs.
- *Quality information* is often difficult to determine. The questions to be asked during the implementation of VLITPs about quality information are: What is quality VLITP information? Where part of the VLITP does it come from? Why is it needed in VLITP? How should it be used in VLITP? Some of these can be answered by using quality standards like: ISO 9000 series, other quality standards; auditing; selecting and managing suppliers.
- *Human aspects of quality* take a historical view of people and their interpretation of quality in their performance. It requires the engaging of employees in assuring quality.
- *Statistics for quality* during the implementation of VLITP often include graphical presentation of data, summary measures, probability, probability distributions and significance testing of the quality of project outcomes.
- *Design and quality* takes into consideration the design and innovation of various aspect of VLITP. They use the systems techniques and methods to design quality in VLITP implementation.
- *Reliability and testing* are carried out with every products and services within a VLITP. They test specifications within sub-projects, the planning phrase of these sub-project as well as the methods, facilities and perceived economic advantages. It is also important to consider the reliability of products and software when modeling and determining specific predictions for VLITP outcomes.
- *Inspection* is carried out regularly based on specific objectives for the inspection. It is very important to know what to inspect and how much inspection to carry out. The manager of VLITP indicates which sampling and inspection methods to use and have measurements of the effectiveness of inspection.
- *Statistical process control* is used for detecting and measuring variation in processes during the implementation of VLITPs. These often found in strategic

project locations with charts for variables, charts for attributes and cumulative sum techniques.

- *Quality improvement* can be indicated by simple or sophisticated ways of improving quality with the aim to generate ideas for improving VLIT quality. It is also used for analyzing failures within a VLITP or as a part of problem-solving methods.
- *Evaluating and looking forward* is the use of strengths and weaknesses as an approach to VLIT quality management. This method is allied to quality management during the various VLIT implementation phrases. It often brings up new ideas and leaves all participants with the prospects for good way of managing VLITP.

Historical Development

Difficulties arise when tracing the history of project management. Some see it as a late modern conceptualization in which case it cannot have a pre-modern history, only harbingers. Others, however, detect project management activities in the pre-modern past. Some writers trace the development of project management-thought back to Sumerian traders and to the builders of the pyramids of ancient Egypt. Slave-owners through the centuries faced the problems of exploiting/motivating a dependent but sometimes unenthusiastic or recalcitrant workforce, but many pre-industrial enterprises, given their small scale, did not feel compelled to face the issues of project management systematically. However, innovations such as the spread of Hindu-Arabic numerals (5th to 15th centuries) and the codification of double-entry book-keeping (around 1494) provided tools for assessing, planning and controlling projects.

Given the scale of most commercial operations and the lack of mechanized record-keeping and recording before the industrial revolution, it made sense for most owners of enterprises in those times to carry out project management functions on a personal basis. But with growing size and complexity of organizations, the split between owners (individuals, industrial dynasties or groups of shareholders) and day-to-day project managers (independent specialists in planning and control) gradually became more common.

19th century: Some argue that modern project management as a discipline began as an offshoot of economics in the 19th century. Classical economists such as Adam Smith (1723-1790) and John Stuart Mill (1806-1873) provided a theoretical background to resource-allocation, production, and pricing issues. About the same time, innovators like Eli Whitney (1765-1825), James Watt (1736-1819), and Matthew Boulton (1728-1809) developed elements of technical production such as standardization, quality-control procedures, cost accounting, interchangeability of parts, and work planning. Many of these aspects of project management existed in the pre-1861 slave-based sector of the US economy. That environment saw 4 million people, as the contemporary usages had it, “managed” in profitable quasi-mass production.

By the late 19th century, marginal economists Alfred Marshall (1842-1924), Léon Walras (1834-1910) and others introduced a new layer of complexity to the theoretical underpinnings of project management. In 1881, Joseph Wharton offered the first tertiary-level course in managing projects.

20th century: By about 1900 project management theories were being placed on a somewhat thoroughly scientific basis (i.e. Henry Towne’s *Science of management* in the 1890s, Frederick Winslow Taylor’s *Scientific management* (1911), Frank and Lillian Gilbreth’s *Applied motion study* (1917), and Henry L. Gantt’s charts (1910s)). J. Duncan wrote the first project management book in 1911 for college education. That was followed by the introduction of Taylorism to Japan in 1912 by Yoichi Ueno—making him the first official project management consultant of the “Japanese-management style” and setting the scene for his son—Ichiro Ueno—to pioneer Japanese style quality-assurance.

The first comprehensive theories of project management appeared around 1920. Henri Fayol (1841-1925) and Alexander Church described the various branches of project management and their inter-relationships. Other followed in early 20th century, including Ordway Tead (1891-1973), Walter Scott and J. Mooney applied the principles of psychology to project management, while other writers, such as Elton Mayo (1880-1949), Mary Parker Follett (1868-1933), Chester Barnard (1886-1961), Max Weber (1864-1920), and Rensis Likert (1903-1981), approached the phenomenon of project management from a sociological perspective.

One of the earliest books on project management--then termed ‘applied management’—was published in 1946 and written by Peter Drucker (1909–2005) called *Concept of the Corporation.* It was the result of Alfred Sloan (General Motors chairman until 1956) commissioning a study of the organisation. Drucker went on to write 39 books, most of which were in the same vein.

Statistical techniques were later introduced into project management studies by the efforts of H. Dodge, Ronald Fisher and Thornton C. Fry in the early 20th century. The 1940s saw Patrick Blackett combined these statistical theories with microeconomic theory and gave birth to the science of operations research. Unlike Taylor’s Scientific Management, this new discipline attempted to take a scientific approach to solving management problems, particularly in the areas of logistics and operations projects. More recent developments into this line of project management studies include the theory of constraints, management by objectives, reengineering, and various information-technology-driven theories such as agile software development, as well as other group management theories.

As the general recognition of project managers' class solidified during the 20th century and gave perceived practitioners of the art/science a certain amount of prestige, so the way opened for popularised systems of project management ideas to peddle their wares. In this context many project management fads may have had more to do with psychology than with scientific theories of managing project activities. Towards the end of the 20th century, project management came to consist of six separate branches, namely:

- Construction Project management
- Manufacturing project management
- Strategic Project management
- Marketing/Advertisement Project management
- Financial Service Project management
- IT Project management responsible for management information systems

21st century–New Emerging Profession: 21st century observers find it increasingly difficult to subdivide project management into functional categories in this way. More and more business processes simultaneously involve several categories. Instead, one tends to think in terms of the various processes, tasks, and objects subject to project management. A number of the areas of project management appear all through the chapters of this book.

Many centuries ago, the career of an accountant fully depended on a specific industry (Reiss, 2007). Accountants today may know nothing or very little about the industry in which they work (i.e. accounting/auditors in hospitals, farms or airline). This allows them to move freely from one industry to another. While project mangers have not quite reached that level of independence, good VLITP managers are rare and may move between industries (i.e. Richard Granger moved from managing London Congestion Charging to National Health Service in the UK). Quite often VLITP managers are selected from those who understand the industry within which the VLITP is implemented. While this may have the advantage of VLIT managers not having to learn what the project object is all about, VLITP management covers issues outside the scope of any one particular industry and requires techniques and strategies that are applicable to many (Reiss, 2007).

In the twenty-first century other factors are beginning to emerge in business decision to implement VLITP in competition to the pursuit of profit—considered by a section of the society to be socially unacceptable (Galliers, 1998; Guah and Currie, 2003). The managers of VLITP must demonstrate positive effect of their strategies and techniques on the host organization, the community at large, the safety of the environment, staff relations and lifestyle.

One consequence is that workplace democracy has become both more common, and more advocated. In some cases, all VLITP management functions are distributing among the project team members—each of whom takes on a portion of the work (Forsberg, Mooz and Cotterman, 2000). These models predate current political issue, and may occur more naturally than does a command hierarchy. All VLITP management, to some degree, embrace democratic principles in the long term. This ensures that members of sub-projects give majority support to overall management; otherwise they'll have to leave and find another job, or in desperate cases, go on strike. Hence VLITP management has started to become less based on the conceptualisation of *classical* military command-and-control, and more about facilitation and support of collaborative activity, utilizing principles such as those of human interaction management to deal with the complexities of human interaction involve in VLITP.

The rise of the Internet and development of collaborative software have instilled a new dimension into VLIT management—the virtual sub-project team (Guah and Currie, 2006). However, VLITP managers know that managing a virtual sub-project team is fundamentally different from managing traditional sub-projects team (Brown, 2001). Virtual sub-project team environments place a heavy weight on project managers to combine their existing mindset, skill and tool in order to manage through technology, rather than simply with technology. The two examples of VLITP, towards the end of this book (see Chapters XI and XII), indicate patterns of effective management of virtual sub-project team management. Several exploratory studies have attempted to address this very issue by combining the collective skills and technologies in VLITP management and collaboration to provide a blueprint for best practices in virtual sub-project environments (Patel and Morris, 1999; Reddy, 2004; Westney, 2001; Youker, 1989).

To exert the positive influence needed for the successful implementation, a VLITP needs people with good project management qualities. Such qualities will be determined, to a large extent, by the circumstances within which the VLITP manager has to exercise authorities. However different situations within VLITP require the display of different qualities for the same manager. While the book considers here the age-old argument of whether a good manager is born or made out of its remit (Daniels et al, 2002; Eyre and Pettinger, 999; Stewart, 1997), there are two elements that are necessary for all good VLITP mangers to understand:

1. VLITP management is essentially always within a team environment where every member of the team has a common purpose; and
2. All members of the sub-project team must accept the authority of a VLITP manager if that manager is to be effective in driving the team to meet the VLITP objectives.

CHAPTER SUMMARY

The field of VLITP management uses several different standards and best practices to increase the likelihood of overall success. Individual standards and practices may vary in complexity and application, but the goals are usually the same-to produce desired project results within the boundaries of time, costs and available resources. Any effective program for VLITP management standards and best practices must provide relevant steps and strategies to guide the selection, management and control of VLITPs. Every VLITP should begin with an approved requirements specification (see Figure 2.2), though requirements rarely present themselves. Before VLITP requirements can be selected and approved, they must be collected, culled and defined.

Managing VLITP cost can be the most complicated, political, (and potentially tedious) element of the project management process. But, costs and expenses have to be controlled, for the sake of IT credibility, and the overall fate of current and future VLITP in the business.

Managing a virtual sub-project is fundamentally different from managing traditional sub-projects. Virtual sub-project team environments place a heavy weight on VLITP managers to combine their existing mindset, skill set and tool set in order to manage through technology, rather than simply with technology which means the decision to implement VLITP in businesses must be supported by the utmost objective of improving profit.

REFERENCES

Angell, I.O., & Smithson, S. (1991). *Information Systems Management: Opportunities and Risks.* Macmillan Press, Basingstoke.

Brown, A. D., & Jones, M. R. (1998). Doomed to Failure: Narratives of Inevitability and Conspiracy in a Failed IS Project. *Organization Studies, 19*(1), 73-88.

Brown, T. (2001, November). Modernization or failure? IT development projects in the UK public sector. Financial Accountability and Management, (17. 4), 363-381.

Currie, W. L. (2003) A knowledge-based risk assessment framework for evaluating Web-enabled application outsourcing projects. *International Journal of Project Management, 21*, 207-217.

Currie, W. L., & Guah, M. W. (2006). Web Services in National Healthcare: The Impact of Public and Private Collaboration. *Information Systems Journal*, 1(2), 48-61.

Currie, W. L., & Willcocks, L. P. (1998). Analysing Four Types of IT Sourcing Decisions. In The Context of Size, Client/Supplier Interdependency And Risk Mitigation. *Information Systems Journal, 8*, 119-143.

Daniels, K., Johnson, G., & deChernatony, L. (2002). Task and institutional influences on managers' mental models of competitions. *Organization Studies*, Jan-Feb.

Eyre, E. C., & Pettinger, R. (1999). Mastering Basic Management. Third Edition, MacMillan Press, London.

Feldman, M. S. (2004). Resources in Emerging Structures and Processes of Change. *Organization Science, 15*(3), 295-309.

Forsberg, K., Mooz, H., & Cotterman, H. (2000). *Visualizing Project Management*, Second Edition, Wiley, pp. 89

Galliers, R. D. (1998). Reflections on BPR, IT and Organisational Change. In *Information Technology and Organisational Transformation: Innovation for the 21st Century Organisation,* (Galliers, R. D. and Baets, W. R. J. eds) Wiley, Chichester, pp. 225-243.

Guah, M. W., & Currie, W.L. (2003). ASP: A Technology and Working Tool for intelligent Enterprises of the 21st Century. In *Intelligent Enterprises of the 21st Century.* Edited by Gupta, JND and Sharma, SK. Summer. IGI Public, pp.188-219.

Guah, M. W., & Currie, W. L. (2006). *Internet Strategy: The Road to Web Services Solution.* Pennsylvania: Idea Group Publishing.

Guah, M. W. (2007). Changing Healthcare Institutions with Large IT Projects. *IRMA Conference 2007,* May 19–23, 2007, Vancouver, Canada.

Kanter, R. M. (2001). The Middle Manager as Innovator. *Harvard Business Review,* September.

Kerzner, H. (1989). Project management: A Systems Approach to Planning, Scheduling, and Controlling, 3rd Edition, Van Nostrand Reinhold, NY, p. 84.

Kaushik K. D., & Cooper, M. (2000, January). *Industrial Marketing Management, 29*(1), 65-83

Patel, M. B., & Morris, P. G. W. (1999). *Guide to the Project Management Body of Knowledge,* Center for Research in the Management of Projects, University of Manchester, UK, p. 52.

Reddy, R. (2004, January). Cultivating the IT Portfolio Manager. *Cutter Consortium Business-IT Strategies Executive Report, 7*(1).

Reiss, G. (2007). Project Management Demystified: Today's tools and techniques. Taylor and Francis, 3rd edition.

Sagripanti, M., Dean, B., Barber, N., (2002). An evaluation of the process-related medication risks for elective surgery patients from pre-operative assessment to discharge. *International Journal of Pharmacy Practice, 10*, 161-70.

Slay, J. (2006). Information Technology Security and Risk Management.

Smith, D. (2000). E-business strategy risk management. *Computer Law and Security Report, 16*(6), 394-396.

Stewart, R. (1997). The Reality of Management. Third Edition, Butterworth-Heinemann, Oxford.

Vincent, C., Taylor-Adams, S. E., & Stanhope, N. (1998). Framework for analysing risk and safety in clinical practice. *British Medical Journal, V.316*, 1154-7.

Westney, R. E. (2001). Risk Management: Maximizing the Probability of Success. *In Project Management for Business Professionals*, edited by Joan Knutson, Wiley, NY, p-128.

Youker, R. (1989). Managing the project cycle for time, cost and quality: Lessons from World Bank experience. *Keynote paper, INTERNET 88, Glasgow, 1988, 7*(1) February, 54.31

Chapter III
Why Implement Very Large IT Projects

ABSTRACT

The basis upon which the objectives and policies for managing a VLITP are formulated is the need to achieve the project objectives on time and under budget. However, benefits for investing in the underpinning activities may not be sufficient to ensure long-term viability for the host organization. This chapter gives a detailed explanation of why very large IT projects are implemented costing the host organization billions of dollars. It also breaks down the management process of VLITPs, giving clarity to procedure and policies.

INTRODUCTION

The implementation of VLITP may demonstrate that profit maximization is a short-term goal of business activity. Investing in activities that affect the society in general show long-term interests of the organization. This explains the alternative view that profit maximization can only be achieved over the extreme long-term participation of the organization. Investing and underpinning VLITP activities sufficiently to ensure long-term viability of organizations can achieve this. These are therefore the basis on which objectives and policies for managing VLITPs are formulated. These are not only the purpose of organizations managing VLITPs but also the means by which their business objectives are to be achieved.

The management team of VLITPs takes seriously their responsibility for the overall direction of the project. This often requires taking strategic decisions, which are then implemented by individual team leaders for sub-projects working in their own particular ways. Members of the management team usually have great access to organization information as well as very good understanding of the value and contribution of various aspects of VLITPs. Work on the management team is better carried out with mutual understanding and mutuality of interest, rather than blind obedience to orders from one individual even if that individual is the chief executive or a well-known project champion.

PROJECT CONCEPTION

Before beginning a VLITP the following goals should be achieve as part of the initial decision to go ahead with the planning stages:

- Obtain top management support for the estimation process often based on an experiential-analogy approach, in order to obtain better adherence of estimated efforts against the actual ones.
- Ensure the necessary skills are available to in order to satisfy the needs of specific aspects of the VLITP, which will in turn improve the quality ranking in the eyes of customers.
- The introduction of standardized and objective techniques, supporting sub-projects by size as the main input for the subsequent effort towards the project estimation process.
- The introduction of a relationship, across different phases of the project, between different techniques.
- The setting up of projects' historical database and ensuring that shared and access are made available to people making initial preparations for the projects.

A VLITP is a collection of logical stages that maps the life of a project from the beginning to the end (see Figure 3.1). This logic is used to define, build and deliver the final outcome of the VLITP. Each stage should provide one or more deliverables, which are needed to move on to the next stage of the project. Deliverables—usually tangible and verifiable outcomes of work that serve to define the work and resources that are needed for each stage—are means of the host organizations evaluating the progress of the project and determining the need to take action to correct errors or mistakes (Cleland, 1990). Figure 3.1 also demonstrates how VLITPs are broken into various stages to make the project more manageable and reduce the risks that are involved with the project. It also provides a better insight on the risks that are involved during the project. Each stage should start when the previous stage ends. Overlapping stages quite often lead to higher risk and should only be done when the risks are acceptable for all the stakeholders involved in the project. At the end of each stage there should be a review of the essential deliverables for that stage.

Define Project Goals

Defining the goal of a VLITP is the first step of the project (see Figure 3.1). The goal focuses on providing value to the host organization. A well-defined goal gives the VLITP team a better vision and a clear focus on the following stages of the project. In general starting projects share the following characteristics:

- The costs and level of staffing is low at the start of a project, this increases as the project is halfway and then decreases when nearing completion.
- Risk and uncertainty are (in general) the highest in the beginning of the project. Once the final goal is defined and the project progresses the risk and uncertainty quite often decreases.
- The ability for stakeholders to change and influence the scope, cost and goals of the project are highest in the beginning of the project. Once the project progresses that ability decreases because the cost of changing the scope and goals increases as the project progresses.

Plan Project

When the goals of a VLITP have been defined a project plan has to be developed (see Figure 3.1). The plan basically answers the following questions:

- What are we going to do?
- Why are we going to do it?
- How are we going to do it?
- Who is going to be involved?
- How much time will it take?
- What are the costs of the project?
- What can go wrong (are the risks and uncertainties), what can we do about it and how can we avoid it?
- How did we estimate the time schedule and project budget (costs)?
- Why were certain decisions made?
- How can we check if the project is successful?

In addition to the above questions for each stage of a VLITP the following issues need to be defined and added in the project plan:

- Deliverables
- Tasks
- Resources
- Time to complete each task.

The project plan is used as a tool to measure the performance of the project throughout the life cycle. It also defines the scope, time schedule and budget of the project.

Execute Project Plan

After the first two stages (Define Project Goal, Project Plan) have been completed it is time to execute the project plan (see Figure 3.1). During this stage it is important to monitor and manage the progress, scope, schedule, budget and people that are working on the project. This needs to be done to ensure that the goal of the project is being achieved. Project changes and progress needs to be documented and compared with the original project plan. Stakeholders need to be involved and committed to the project. At the end of this stage the project team delivers a finished product to the organization.

Figure 3.1. Generic project life cycle

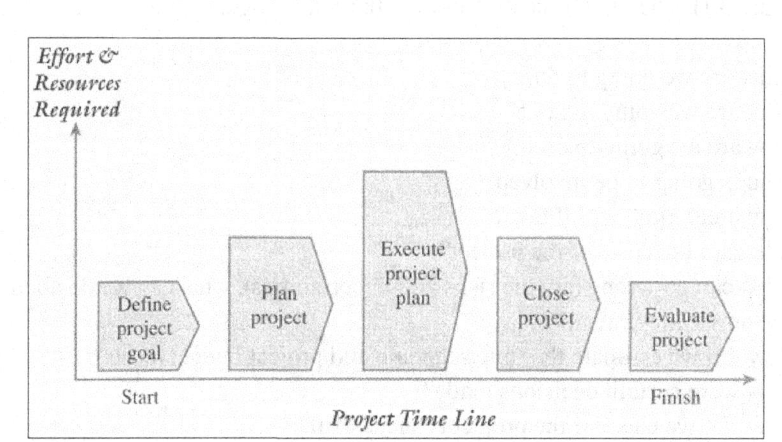

Close Project

Under normal circumstances a project is considered closed when the project team has delivered a finished product. Unlike normal projects, VLITPs do not usually deliver a finished product. Instead VLITPs are officially considered closed when the host organization has agreed that a reasonable outcome has been delivered in consideration of the expenditure at a particular point in time. This is due to the fact that VLITPs normally face the problem of continuous changes to objectives. It quite often means escalating costs for the project. This stage ties up all the loose ends of the project and checks if all the important deliverables have actually been delivered. It is also a formal acknowledgement by the host organization that the desired products have been delivered in an acceptable form. When this happens, the closure of a VLITP occurs with a final project report and presentation for the by the project management team to the host organization.

Evaluate Project

After a formal acknowledgement stage VLITP goes through an evaluation process investigating the entire project. The evaluation VLITPs can be carried out using a combination of several methods due to their size and complexity. Some parts of a VLITP are evaluated using the return-on-investment (RIO) method to measure the impact of the project for the host organization. Yet others can be evaluated by documenting the experiences gained by the host organization during the implementation of a VLITP—lessons learnt which could improve the future performance.

Issues investigated during this phase of VLITPs involve:

- Identifying which aspects of the project were accomplished brilliantly.
- Identifying which aspects of the project could have been improved.
- How many unnecessary risks were undertaken.
- Which outcomes of the project are still venerable of going wrong in the future.
- Which aspects of the project could still be improved and at what costs.
- Whether the final solution the intended outcome or did it happen as a by-product.

The purpose of this exercise is to document the "lessons learned" from the VLITP often compiled in a template that can be shared within the host organization. The report may subsequently become best practices because of its use.

In addition to evaluating the outcomes and processes of VLITP, the performance of all the sub-project managers should be evaluated along with the overall project manager. The support of all members of the VLITP is needed in the form of giving honest feedback not only on the projects but also on the performance and personal goals of each other during the entire project. In most cases a third party company would be contracted to do a complete audit of the VLITP.

PROJECT COST

Despite the vast cost and number of years it takes to complete—in a century of global competition, tightened budgets, and downsizing—VLITPs are readily terminated, even when they are perceived as being late or over budget. But, where the host organization understands the expected benefits of the project and is committed to achieving its objectives, the VLITP stands a better chance of continued funding.

Accurate identification of the following variables with greatly helps in determining the overall costs for a VLITP:

- **Limitation to project scope:** VLITPs often expand in scope as the idea gets sold both internally and externally to an organization.
- **Project escalation procedures:** It is important to determine which stakeholders will make the final decision about certain aspects of the project become doubtful.
- **Work breakdown structures:** While it may not always be possible to have a detail breakdown of work required for a VLITP, managers of sub-projects should be required to develop a detailed task list for anticipated objectives.
- **Time estimate:** After compiling the above work breakdown structure from various sub-projects, the project planners can estimate the length of time required for a VLITP.
- **Project management flow charts:** The development of a flow chart for VLITP management can prove very useful in successfully managing a VLITP.
- **Resource identification:** The required resources need to be identified and budget for as accurately as possible.
- **Risks identification and evaluation:** The identification of risks is necessary to prepare contingency plans. Also necessary is the evaluation of risks at various stages of a VLITP.
- **Identification of interdependencies:** Many interdependencies can be found in a VLITP simply due to the complexity and time involved. Every time the scope changes an addition is made to the interdependency in a VLITP.
- **Identify and track critical milestones:** A critical milestone is needed to determine if a VLITP is going according to plan. Quite often the identification process is full of ambiguities due to interdependencies mentioned above.

When the critical milestones have been identified, it is necessary to track them through the entire project.

- **Project phase review:** This aspect of an IT project is quite often ignored for most standard project but very important for a VLITP however costly it may seem. This additional cost is less then the cost of project failure.
- **Securing necessary resources:** When items of VLITPs are competing for budgetary allocations, managers of sub-projects must ensure much needed resources are secured.
- **Change control:** This is often considered an unnecessary administrative expenditure. However, the process managing change control is vital to ensure VLITPs keep on track and within scope.
- **Number of vendors:** VLITPs have to follow the government rules of open contract bidding systems. This often leads to the sub-contracting of work on the project to different vendors. Considering all participating businesses would need to make profit, the number of vendors involved in the project would finally determine the total costs of a VLITP.
- **Number of sub-projects:** This aspect of a VLITP is determined by its scope. The number of sub-project determines the number of teams, which in turn determines the total staff required for the entire project.
- **Number of systems connection:** This aspect is also determined by the scope of the project and the total number of sub-objectives a VLITP has been broken into.
- **Project length:** In general the longer a VLITP lasts the more it costs—an aspect greatly influenced by project escalation mentioned above.
- **New technology:** Because VLITPs are expected to delivery one or more emerging technologies as part of the final outcomes, they quite frequently incorporate the latest technologies within the particular sector or industry. This adds to the costs of implementing VLITPs because new technologies are often more expensive then older ones.
- **Expertise availability:** The human resource costs of VLITPs are influenced by the availability of necessary expertise in the locality. In most cases, VLITPs have to depend on expatriates for the necessary expertise not found nationally.

Several researchers have shown that things work better when VLITPs have a "champion" in upper management ranks (Arthur, 2000; Chin et al, 2004; Cleland, 1990; Patel and Morris, 1999). This champion often sees to it that the project needs are not only understood by the top executives of the host organization but also totally convinced about the value of this project to the host organization. In cases where a willing "champion" for the VLITP cannot be found among top executives—usually as a result of major change in top management—it can be interpreted as a sign for a complete review of the entire idea of this project. By reassessing the commitment from the host organization, VLITP managers have another opportunity to sell the anticipated benefits to the host organization. Likewise the host organization could decide to stop funding the VLITP at certain stage and absorb the losses so far.

During the beginning of the 21st century, the U.S. Government spent more than $38 billion each year on IT—a number that has continue to grow as virtually all functions of Government take advantage of efficiencies provided by IT (Arthur, 2000). In cases where the VLITPs are well selected, controlled, and managed, agencies are provided with some of the best opportunities to fulfill their missions with the lowest cost and greatest benefits. The most publicised among these projects was Y2K. This included a priority management objective to strengthen a government-wide management strategy by using capital planning and investment control to better manage VLITPs. Most Y2K strategies focused on the Government's process of deciding levels of VLITP investments for various programs as well as the benefits of programs and performance benefits.

FITTING IN AN INSTITUTIONAL SETTING

The type of institutional structure of organizations implementing VLITPs can have significant effect on the project management strategy (Weill, 1993). Eyre and Pettinger (1999: 93) refer to this as 'unity of command'. While chief executives are not required to act entirely on their own when setting objectives for VLITPs, the top management needs to speak with one voice. Where applicable, they should reconcile their differences before strategic directions are given to the management team of a VLITP. Similarly individual project managers on VLITPs should not be allowed to express adverse opinions about instructions that have to be implemented. Such actions may lead to confusion as a result of project staffs being required to follow directives from numerous sources (Morris, 1998).

Inherent in this institutional structure format is the need to make decisions because management teams of all VLITPs are responsible for decision-making—an important aspect of their capacity to delegate responsibilities to members of the project teams (Barney and Griffin, 1992; Chin, Brown and Hu, 1992; Feldman, 2004). An environment with inappropriate institutional structure often leads to lack of courage or lack of self-confidence by managers of sub-projects and team leaders to make decision or even refrain from taking actions that may affect various aspects of VLITPs.

A committee of managers from the host organization normally decides on the need to implement a VLITP with vested interests in vary aspects of the projects.

These managers are expected to have specific expertise that would facilitate effective monitoring of the performance of the project management team. Such monitoring committee needs to be set up—either by appointment or election depending upon the particular circumstances—to deal with specific areas of the project activities. The monitoring committee must have clearly defined authorities and functions, with clearly drawn up rule and guidelines. Like all other aspect of business management, monitoring committees of VLITPs should seriously consider the following to ensure successful project implementation (Morris, 1998):

- **Authority:** This is where the monitoring committee has duties and authority given to it to ensure that those duties are carried out. The precise extend of such authority must be clearly defined with limitations. This obliges the monitoring committee to be accountable for its actions as well as the quality and effectiveness of the tasks and duties it performs.
- **Duty:** This is where the monitoring committee is obliged to comply with certain instructions and limitations given it by those the chief executive (or top management) of the host organization.
- **Responsibility:** This is where the monitoring committee ensures that duties are properly, adequately and effectively carried out. This may entail that the monitoring committee may delegate authority and accountability. Not only is responsibility a key part of all effective management of VLITPs, the ultimate responsibility for the effective completion and successful achievement of project results rests with the chief executive (or top management) in the host organization.
- **Accountability:** This is where the monitoring committee reflects the acceptance of specific responsibilities for tasks, duties and actions. It enables people to be called to account for the quality and volume of their work as well as performance of their teams, units and divisions. Monitoring committees can assist chief executives of host organizations to account for the performance of a VLITP.

The above can be compared to the undisputed truth of European colonization in Americas, Asia and Africa—the state played a central role in colonial enterprises. European colonies emerged out of trading ventures organized as joint stock companies chartered by colonizing states in which the crown invested both its prestige and its capital. Colonial territories were conquered and defended by soldiers and sailors paid either by the colonizing state or the local colonial state. Plantations and mines were often directly owned by the local colonial states. The truth about "winning" the western frontiers of the United States and Canada makes for even poorer libertarian dramas (Mack, 1974).

A better historical analogy for VLITP is provided by one of the greatest civil engineering project of last century—the construction of the Panama Canal (Mack, 1974). Like any VLITP, constructing the Panama Canal required overcoming tough new engineering and science problems in an unforgiving environment. A labor force had to be imported and supported, and sufficient capital had to be invested despite the fact that private investors could not provide the financing necessary to complete the tasks. After twenty years of failed efforts by private French firms to dig a canal across the Isthmus of Panama and the failure of a private American firm to dig a canal through Nicaragua, it was the United States government that successfully completed the construction of the Panama Canal (Mack, 1974). Financed by the United States government and management by U.S. Army engineers succeeded where the private sector failed. Even more difficult problems than those encountered in constructing the Suez Canal, were eradication of yellow fever and the control of malaria in Europe (Mack, 1974).

CUTTING ACROSS FUNCTIONAL LINES

The implementation of VLITPs entails co-ordination, involving the integration and synchronisation of various business activities in the host organization. Coordinating all functions often leads to harmonious work towards common objectives. Co-ordination of VLITP activities and functions is clearly of utmost importance to the project's successfulness whether the functions are interdependent or not.

Much of how VLITPs co-ordinate activities in the host organization depend on the existence of goodwill between the various departments in the organization. This requires top executives to make genuine effects in promoting good relations between various units, divisional managers and staff. Although certain management styles in the 21st century foster a competitive spirit between different departments for the purpose of generating aggressive enthusiasm and showing superiority amongst departments and units, they often end up with bad feelings between the various departments—usually a result of disregarding the needs of other departments, creating some form of imbalance in the organization's business goals as well as a breakdown in co-ordination (Rodrigues and Hickson, 1995). VLITPs need to co-ordinate one division with another while controlling failure risks. Setting out clear objectives and goals for VLITPs ensures good understanding and appreciation amongst managers and staff from all divisions. The precise nature of requirements and expectations must be explained together with responsibilities, authorities and duties of all involved.

Few practical actions are required by the management team of VLITPs to improve co-ordination within the host organization. These activities begin with having the strategy, objectives, procedure and policies clearly defined and explained to all involved. These should include clearly defined procedures for each activity to check adherence to project requirements—co-ordination is likely to fail without some form of control mechanism. All member of VLITP teams need to understand how their activities fit into the whole effort of the host organization to achieve certain objectives. This is done either by induction when new staffs are appointed or by arranging formal and informal update meetings for all staff. This means of communicating the overall objectives of VLITPs can only be effective if they are opened and honest (Patel and Morris, 1999). Committees, cluster groups, and various sub-projects are also useful in promoting effectiveness of co-ordination due to the primary reason of breaking down barriers between departments, units and functional divisions. They also enable mutual confidence and trust to be engendered with a greater understanding of the pressures and constraints that are often felt when VLITPs are being implemented.

REFLECTING ON WIDER IMPLICATIONS

VLITPs can be understood as massive public interest activities, implemented to meet the demands, needs and economic requirements for the general public. If there are new advancement resulting from the implementation of VLITPs, they will attract and keep the kind of customers the hosts organization seeks, thus adding value to customers. Making the host organization's products and services more attractive to the general public will provide plenty of scope for private firms to profit from the provision of goods and services. Some private firms are unwilling to take the necessary risks to finance VLITPs implementation from which such an attraction could cultivate.

VLITPs require an effort both on the business side (that has to adapt its requirements to a scope defined by business efficiency) and the technological aspects (that has to adopt a more business minded attitude). Their failure is by no means a seldom or an occasional issue. It is widespread and all pervading as the statistics over VLITP failure rate testifies (Turnbull, 2003). A deeper look into the casualty records provides an appreciable insight into the causes of VLITP failure.

The proper management of VLITP is therefore a pragmatic approach that helps businesses deliver and realize the required benefits, innovation, and new ways of working intended to takes them through this century. In this century change is a way of life for all businesses. The outcomes of a successful VLITP are the delivery of new and improved services, introduction of new processes, change in relationships with supplier, and organizations that merge and divide in response to political or market forces. This is part of businesses new effort today to strive and achieve excellence by improving practices and services, to be better prepared for the future, to make innovation possible and to encourage new ways of thinking about doing business.

Where a VLITP leads to a major change there is normally complexity and risks, and many interdependencies to manage and conflicting priorities to resolve. Such projects impact business policies and major management elements including:

- **Investment management:** The management of IT investment is an integrated view that provides a good way of managing life cycle of VLITPs. This involves three phases of the IT management process (selection, control, and evaluation). The selection phase is where the host organization determines priorities and makes decisions about which VLITPs can be funded based on the technical soundness, anticipated contribution to business needs, performance improvement priorities, and the overall IT capital funding levels. The costs, benefits, and risks of all VLITPs are assessed and different projects are compared against each other resulting to ranking by priorities. In the control phase, various sub-projects are compared at similar stages in development. There is a review of progress looking at projected cost, schedule, and anticipated benefits—conducted at key milestones in each project's lifecycle. This is followed by an evaluation phase comparing actual performance against estimates to identify and assess areas of the projects in which future decision-making can be improved.
- **Information management:** This is where VLITP managers consider the collection, use, and dissemination of information contained in a VLITP. Each system should ensure public access to records where required and appropriate. The system involved should be designed to collect or create only that information necessary for the proper performance of business functions and has practical utility. VLITP managers must ensure that proper records are maintained, and that the public is not unduly burdened by the requirements to provide information mandated by an internal unit.
- **IT architecture:** This consists of logical and technical components that funnel the development and evolution of a collection of related systems. Chapter X details SOA that provides a high-level description of business's goals, the departmental functions being performed, and the relationship among such functions, the information needed to perform the tasks, and the flow of information among the business units. SOA provides the rules and standards needed to ensure that interrelated sub-projects are built to be interoperable and maintainable (Marks and Bell, 2006).
- **System development environment:** Every VLITP needs an environment that is good for systems development. Application of varies elements in that environ-

ment to system development can provide consistent management and control of VLITPs. That's because systems development is the most sensitive part of VLITP—where most failures occur. Where systems that are being developed do not meet expectations or are never used, the outcomes of a VLITP would stand no chance of being made operational. A good environment is needed for VLITPs to develop software using project's staff, use a contractor to develop software, use commercial off-the-shelf software, or even a combination. To effectively manage software development and acquisition processes, however, VLITP mangers need to have well defined software management processes, including methodologies and standards that will be used.

- **Information security:** Policies and practices for securing information in VLITP provide the framework to protect the host organization's IS resources and assets. Such protection ensures the integrity, appropriate confidentiality, and availability of the data for VLITPs. Integrity ensures that data have not been altered or destroyed in an unauthorized manner. Confidentiality ensures that project information is not made available or disclosed to unauthorized individuals or entities within or outside the VLITP. Availability ensures that data will be accessible or usable upon demand by an authorized entity with regards to a particular VLITP. Key activities for managing information security in VLITPs are risk assessment, awareness, controls, evaluation, and central management. Risk assessments consist of identifying threats and vulnerabilities to information assets and operational capabilities, ranking risk exposures, and identifying cost-effective controls. Awareness involves promoting knowledge of security risks and educating users about security policies, procedures and responsibilities. Evaluation of VLITP involves monitoring effectiveness of controls and awareness activities through periodic evaluations of sub-projects.

KEEPING THE PRIMARY FOCUS

Regardless of the number of technologies that emerges from a particular VLITP implementation, critics and other observers will point out failure in achieving the primary objectives set by the host organization—something referred to in strategy literature as *Strategy and Structure* (Chandler, 1962). It emphasises the need for staff on VLITPs to understand what has to be done for successful implementation. Quite often in VLITPs, subordinates misunderstand instructions from the project managers. That frequently results to lack of staff motivation leaving unhelpful staffs that only see difficulties in enlisting cooperation. One common problem during the implementation of VLITPs is the lack of wisdom from certain managers to acknowledge the limitations of their own authorities. As far as possible, VLITP managers should avoid weakening their authorities by trying to exercise them where they are likely to be challenged or ignored. All VLITP managers should acknowledge that persistence, drive and political manipulation—supported by an understanding of colleagues' motives and a correct assessment of the political situation—are necessary to get the project plans successfully implemented.

Project failure is not only defined by objective criteria but also by the *perception* of the respondents. The advantage of a perception is that it naturally integrates multiple aspects. Perception, on the other hand, can be inevitably partial. The reality of an individual's perception can inevitably be embellished in cases where the respondent has taken an active role in the project. Unlike a project that had been imposed on staff of the host organization, many members of staff would cast a grimmer look at the project's outcome. Such insight into project perception quite often challenges the VLITP management teams to engage in what is frequently referred to as "ontological designing". This is the self-conscious construction of technological worlds supporting a desirable conception of what would be humanly acceptable in a thriving business environment.

CHAPTER SUMMARY

The chief executives of host organizations are responsible for taking the final decision to implement VLITPs—thus, making them ultimately responsible for the successful achievement of objectives and long-term efficiency of the organization. For this reason, all chief executives should understand the process, and have confidence in the management team of VLITPs. The planning stages of VLITPs cannot be over-emphasized because those responsible may just be deciding on the future of the entire organization. Their plans for VLITPs contain both long-term and short-term objectives of the organization, assessing the opportunities and consequences that may arise in the future. These in turn establish the basis for short-term and long-term successful or failure in the organization's activities.

As a key project management activity co-ordination is a key responsibility of management team of VLITPs. The contribution of co-ordination is clearly a key management task and activity that benefits departmental and corporate activities in the host organization. It demonstrates the combined and co-ordinated efforts of all units and divisions that contribute to the effectiveness and maximization of the host organization chances of achieving its long-term objectives. This chapter has challenged such clichés by reconceptualizing the relation of VLITPs, rationality, and modernization.

The chapter has also argued that however desirable an alternative goal may be, no fundamental change will take place in the host organization as a result of VLITPs that ignores the business environment and consequently sacrifice millions of staff. The essence of VLITP is to promote the value of individuals in a well-run organization taking advantage of all the modern technologies available to the host organization.

REFERENCES

Arthur, J. (2000). Management of Large Public Information Technology Projects. *Presentation at USDA Forest Service.* (www.policyworks.gov/policydocs /policy_list.htm).

Barney, J. B., & Griffin, R. W. (1992). *The Management of Organizations: Strategy, Structure, Behavior.* Houghton Mifflin, Boston.

Chin, P. O., Brown, G. A., & Hu, Q. (2004). The Impact of Mergers and Acquisitions on It Governance Structures: A Case Study. *Journal Of Global Information Management, 12*(4), 50-74.

Cleland, D. I. (1990). Project Management: Strategic Design and Implementation. TAB Books, PA, p. 23.

Eyre, E. C., & Pettinger, R. (1999). *Mastering Basic Management.* Third Edition. MacMillan Press, London.

Feldman, M. S. (2004). Resources in Emerging Structures and Processes of Change. *Organization Science, 15*(3), 295-309.

Guah, M. W. (2008). Changing Healthcare Institutions with Large Information Technology Project. *Journal of Information Technology Research, 1*(1), 14-26.

Mack, G. (1974). T*he Land Divided: A History of The Panama Canal and Other Isthmian Projects.* New York: Octagon Books.

Marks, E. A., & Bell, M. (2006). *Service-Oriented Architecture: A Planning and Implementation Guide for Business and Technology.* New Jersey: John Wiley and Sons.

Morris, P. W. G. (1998). Key Issues in Project Management. *In Project Management Handbook* edited by J. K. Pinto, Jossey-Bass, p-5.

Patel, M. B., & Morris, P. G. W. (1999). Guide to the Project Management Body of Knowledge. *Center for Research in the Management of Projects*, University of Manchester, UK, p. 52.

Rodrigues, S. B., & Hickson, D. J. (1995). Success in Decision Making: Different Organizations, Differing Reasons for Success. *Journal of Management Studies*, 32(5), 249-678.

Turnbull, A. (2003). Managing Successful Programmes by the Office of Government Commerce (OGC). HM Stationery Office: UK, (2nd Edition), Foreword, p. 5.

Weill, P. (1993). The Role and Value of IT Infrastructure: Some Empirical Observations. In *Strategic Information Technology Management: Perspectives on Organizational Growth and Competitive Advantage*. M. Khosrow-Pour and Mahmood, M. (eds.). Hershey: Idea Group Publishing, pp. 547-72.

Chapter IV
Methodologies for Implementing VLITPs

ABSTRACT

Different VLITP methodologies are capable of solving various types of problems during a project life cycle. This chapter shows that effect of VLITP methodologies can be widespread, especially in regards to project phases, resource allocation, project monitoring, adjustment of project scope, correction activities, and so forth. It reviews several methodologies that are often used to implement VLITP.

INTRODUCTION

Project methodologies allow businesses to maximize the value of VLITPs for themselves—usually by changing focus. Charvat (2003) describes project methodology not only as a mindset used by businesses to reshape their entire organizational processes, but also as a radical cultural shift for organizations. The problem with this description is the need for methodologies to change as businesses do. With this clarification, a VLITP methodology can be defined, as a list of activities to do that can be adapted to a particular situation, within a specific period of time. Such a list would control and lead the actions of all members of VLITP teams during the life of the project. This requires that all members of the team be familiar with the project methodology, support and use it throughout the life of the project.

There are times when project methodology can go wrong (Cleland, 1990; Patel and Morris, 1999). During such times the VLITP managers need to be able to identify which aspects of the methodology is not working or those resulting to inappropriate resolutions. Charvat (2003) suggests the following project methodologies shortcomings:

- They are often rather abstract and presented at high level than project managers can deal with.
- They sometimes contain insufficient narratives to support problems resulting from their usage and may not have any performance metrics.
- They could either be non-functional or incapable of addressing problems with crucial areas of the project.
- They use non-standard project conventions and terminology, ignoring the industry standards and recognized best practices.
- They may look impressive but frequently lack real integration into many business issues of real importance to the host organization.
- They compete for similar resources without addressing this problem.
- They could even take too long to complete because of bureaucracy and administration.

REVIEW OF EXISTING METHODOLOGIES

Each year, the largest companies in the world cancel hundreds—even thousands—of VLITPs before completion. These projects might be new application software (custom or off-the-shelf), new network architectures, or even new desktop operating systems, but they share one characteristic: discontinued by management before completion. The host organization never reaps the anticipated benefits and wastes thousands, and sometimes millions, of dollars (Keil, 2000). This section examines some of the primary reasons that VLITPs fail, and outlines some of the steps that can be taken to avoid these pitfalls and give VLITPs their best chances for success.

A VLITP is a series of steps, completed by team of project managers over a very long period of time to achieve that goal. Therefore, developing the ability to plan and complete a VLITP is critical to reaching the direction of the host organization's goals and objectives (Rosenau and Githens, 2005). Managing VLITP effectively is one of the most important capacities an organization can possess, as these projects comprise the work of many people, and of any new product, service or initiative (Tao, 2000).

The following are many ways the host organization can ensure a VLITP will reach completion stage:

- **Easier project planning:** When people in the host organization understand the big picture of managing VLITP, they will be able to start on the right principles. Often, it seems getting started is the hardest part of all VLITPs. Starting on the right principles ensures the VLITPs takes off right and subsequently heads towards success.
- **Using the right tools:** There are many approaches, methods and tools that can help the project management team to ensure the project is more effectively managed. While some tools may contain very complex approaches, the focus should be on best practices and tools applicable to both complex and more straightforward aspects of VLITPs.
- **Greater confidence:** The management team will learn new skills and techniques each day as the project progresses during the various stages (see Project Life Cycle) of implementation. But something else is happening too. As they learn and apply these skills and techniques, their confidence in implementing VLITPs will be growing. They will be gaining confidence as the project progresses which contributes to their desire to succeed with the current projects and even prepare them to for future projects.
- **Finish faster:** VLITPs take a lot of time. Managing them effectively can take massive amount of time and effort. Getting the right people on the project management team with the right set of skills often reduces the amount of time it will take to achieve the final outcomes of a VLITP.

Project Life Cycle

The Project Life Cycle (PLC) is a collection of logical stages that maps the life of a project—from beginning to end. VLITPs are broken into various stages to make the project more manageable and to reduce the risks that are involved with the project. These stages enables project managers to define, build and deliver the products of a VLITP. Each stage should provide one or more deliverables needed to move on to the next stage. Deliverables are tangible and verifiable products of work that serve to define the work and resources needed for each stage of a VLITP. At the end of each stage there should be a review of the essential deliverables for that stage. Doing this allows the hosts organization to evaluate the progress of a VLITP and to take timely action to correct errors or mistakes and avoid potential serious delays to completions. This also provides a better insight of the risks involved during the project by verifying the level of ease or complexity at which each stage starts when the previous stage ends. Overlapping stages lead to higher risk and should only be done when the risks are acceptable for all the stakeholders of the VLITP (Cleland, 1990).

Figure 4.1. Systems development life cycle

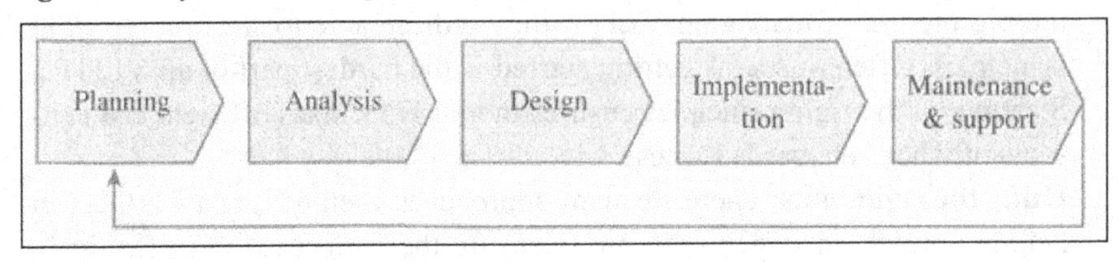

Systems Development Life Cycle

The PLC for many VLITPs is the Systems Development Life Cycle (SDLC) representing the sequential stages an IS follows throughout its life span (see Figure 4.1). While a generally accepted SDLC does not exist, Figure 4.1 shows a number of generally accepted stages associated with system development as used in VLITPs.

Planning: This stage begins by the identification of a specific problem or opportunity that requires system development processes and activities. This stage ensures that the goal, scope, budget and time schedule are based on the most accurate assumptions possible. It also verifies that the most appropriate technology, system development processes, methods, tools are in place for the VLITP to be implemented successfully.

Analysis: Here is where a detailed analysis is made for the problem or opportunity identified the planning stage (see Figure 4.1). Such analysis is based on an good understanding of the current situation by speaking to different stakeholders of the system (users, managers, customers) for clarifications about the problem or opportunity. The data gathered from such meetings helps to form the requirements for the future system.

Design: During this stage of VLITPs, data collected from the previous stage (see Figure 4.1) is used to design the architecture to support the new IS. Such architecture often consists of designing the network, hardware configuration, databases, user interface, application programs and software configuration (Rosenau and Githens, 2005).

Implementation: Following the design stage of the VLITP, the new system is considered ready for implementation. This includes the development (or construction), testing and installation of the system. This stage of a VLITP also includes the user training, support and documentation for the new system.

Maintenance and Support: Although some may not consider maintenance and support a true stage of SDLC, their importance to factors that affect the successfulness of VLITPs demands inclusion. Any changes required to the new system after it has been implemented either falls under maintenance or support functions of the project. Such changes could be actual maintenance of work involved in the previous stages of the VLITP, added functionality, or even fixing of earlier errors. Support, on the other hand, could be delivered in the form of a call center or help desk.

PLC vs. SDLC

Although PLC and SDLC have numerous differences, they also share a lot of similarities (see Figure 4.2). The main difference in the two life cycles is the focus of each. The PLC focuses on the processes of managing a VLITP while SDLC focuses on creating and implementing a product or service. Figure 4.2 shows a comparison of the two popular life cycles used in VLITPs today.

PRINCE2

Managing VLITPs involves proficiency in the use of methodologies such as PRojects IN Controlled Environments version 2 (commonly known as PRINCE2 and increasingly becoming widespread as UK government standard project management tool) or an equivalent. This section looks at the ability to apply an appropriate subset of these methodologies. The advantages of PRINCE2 includes (Wideman, 2003):

- The capacity to completely clarifies people's roles in VLITPs;
- Ensuring that lines of communication are clear;

Figure 4.2. PLC vs. SDLC

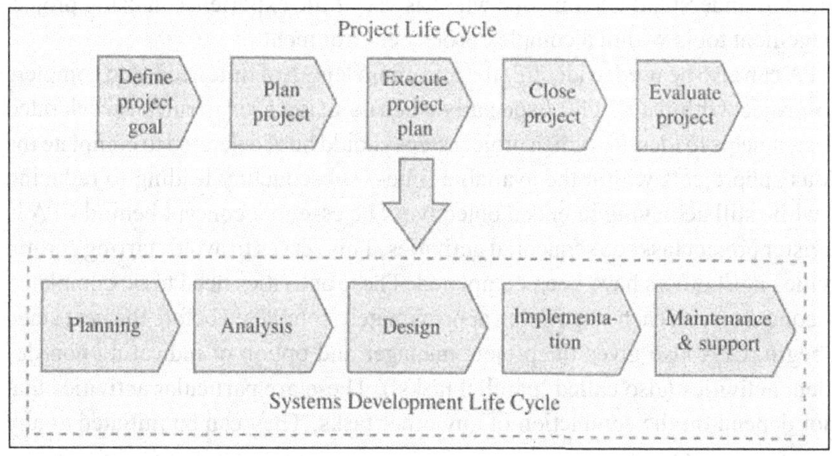

- Making sure that project risk is actively managed;
- Setting up appropriate controls;
- Establishing baseline costs;
- Showing consistency in schedules of various sub-projects;
- Setting scopes for various sub-projects.

PRINCE2 does not only embody project activities but also codifies much of the best practices in project management (OGC, 2003).

Critical Path Analysis and PERT

Another set of practical tools that can be used to aid in the scheduling and management of complex subsets of VLITPs is Critical Path Analysis (CPA) and PERT. These are powerful tools that have been constantly developed to incorporate experiences in managing sub-projects of VLITP since they were developed in the 1950s to control large projects for the defense sector. CPA is very useful for planning all tasks that must be completed as part of a particular subset of a VLITP. It acts as the basis not only for the preparation of a project schedule, but also for resource planning. While managing a VLITP, CPA allows the project manager to monitor achievement of indicated goals. This intends highlight the remedial action to be taken for the project to get back on track.

In addition to the accurate identification of tasks that must be completed on time for the VLITP to be delivered on schedule, CPA also identifies which tasks can be delayed for a while in situation where resource may need to be reallocated to catch up on missed tasks of sub-projects. Notwithstanding, CPA has a major disadvantage in that the relation of tasks to time is not as immediately obvious as with other common tools (i.e. Gantt Charts). This can result in making CPA more difficult to understand for someone who has very little experience in using project management tools within a complex project environment.

CPA can also be used to identify the minimum length of time needed to complete a sub-project within a VLITP. Especially when there is a need to run an accelerated project, it helps to identify which project steps should be accelerated to complete the various subprojects within the available time—subsequently leading to reducing cost while still achieving intended objective. The essential concept behind CPA is to register project tasks as sequential activities. This serves to avoid starting certain activities until others have been completed. These activities need to be completed in a sequence, with each stage being approximately completed before the next stage can begin. CPA also gives the project manager and option of indicating non-dependent activities (also called 'parallel tasks'). These are particular activities that do not depend on the completion of any other tasks. They can be initiated at any time before or after a particular stage is reached. Table 4.1 shows the steps taken to draw a CPA chart where each activity shows the earliest start date, estimated length time required, and also indicating type as parallel or sequential (sequential tasks show which dependent stage). The availability of resources determines the start week and a particular task can be categorized as parallel or sequential depending largely on context.

Another advantage of CPA is the opportunity to plot various activities on a chart using circle and arrow diagrams. These circles show events within the project, such as the start and finish of tasks and therefore numbered to allow easy identification. An arrow running between two event circles shows the activity needed to complete that task. A description of the task is written underneath the arrow. The length of the task is shown above it. By convention, all arrows run left to right. In situations where one activity cannot start until another has been completed, the arrow for the dependent activity begins at the completion event circle of the previous activity. This makes it easy to identify all the activities that will take place as part of a subproject within a VLITP. The aim is that certain tasks must be started and completed on time if the overall project is to be completed on target. This is the 'critical path' - these

Table 4.1. Task list: Implementation of hospital information systems

Task	Possible start	Length	Type	Dependent Variable(s)
1. High level analysis of healthcare	Month 1	5 Months	Sequential	
2. Selection of hardware platform	Month 1	1 Weeks	Sequential	1
3. Installation and commissioning of hardware	Month 3	2 Months	Parallel	2
4. Detailed analysis of Major systems	Month 1	2 Months	Sequential	1
5. Detailed analysis of component parts of health-care information	Month 1	2 Months	Sequential	4
6. Programming of Major systems	Month 4	3 Months	Sequential	4
7. Programming of component parts	Month 4	3 Months	Sequential	5
8. Quality assurance of Major systems	Month 5	1 Months	Sequential	6
9. Quality assurance of component parts	Month 5	1 Months	Sequential	7
10. Training on Major systems	Month 7	1 Week	Parallel	6
11. Compilation of implementation reporting	Month 6	1 Month	Parallel	5
12. Compilation of System Maintenance procedure	Month 6	1 Month	Parallel	5
13. Analysis of Systems Maintenance procedure	Month 6	2 Months	Sequential	5
14. Detailed training	Month 7	1 Month	Sequential	1-13
15. Final Project Presentation	Month 4	2 Months	Parallel	13

activities must be very closely managed to ensure that activities are completed on time. The tool advises VLITP managers to take immediate action if jobs on the critical path slip in order to get the project back on schedule. Thus avoiding a situation where the completion of a VLITP will slip.

There are often instances when the VLITP mangers may need to complete a sub-project earlier than the original CPA says is possible. In such cases the VLITP mangers need to take action to reduce the length of time spent on VLITP stages. Careful attention should however be given to resource planning. It is easy to pile resources into every project activity to bring down time spent on each. On the other hand, this is likely to consume huge additional resources. The common example is a quick decision to use double the number of staff on a particular subproject because a need has arisen to complete it few months earlier. This could have the overall advantage of shorten a VLITP by several months or even a year, but would normally raise the overall cost of a VLITP.

PERT is a variant of CPA that takes a more sceptical view of the time needed to complete each stage of a VLITP. It takes a slightly more skeptical view of time estimates made for each project stage (see Figure 4.2). It estimates the shortest possible time each activity will take, the most likely length of time, and the longest time that might be taken if the activity takes longer than expected. The below formula is used to calculate the time to use for each project stage: (Shortest time + 4 x likely time + longest time)/6.

This formula is commonly used in VLITPs to help the bias time estimates away from the unrealistically short time-scales normally assumed.

The following points are therefore key to ensure VLITP managers take advantage of CPA as an effective and powerful method of assessing VLITPs:

- What tasks must be carried out?
- Where parallel activity can be performed?
- The shortest time in which you can complete a project.
- Resources needed to execute a project.
- The sequence of activities, scheduling and timings involved.
- Task priorities.
- The most efficient way of shortening time on urgent projects.

While CPA can be very useful for assessing the importance of problems faced during the implementation of VLITP plans, an effective CPA can make the difference between success and failure of VLITPs due to their complex nature.

SOFTWARE DEVELOPMENT MODELS

This section discusses the various software development models used in VLITPs. It will focus on Waterfall model and Iterative model, the two principle software development models upon which various software development frameworks have been developed in the past. Thus it is necessary to discuss these two models before looking into the various software development frameworks.

Waterfall Model

The waterfall model is a sequential software development model (developed by W. Royce) wherein software development is seen as steadily downwards flowing process (like a waterfall) through the phases of software development as in Figure 4.3 (Waterfall Model, 2007).

In Royce's waterfall model, the following phases are followed perfectly in order:

- Requirements specification
- Design
- Implementation (Construction)
- Integration
- Testing and debugging (also know as verification)
- Installation
- Maintenance

Figure 4.3. Waterfall model

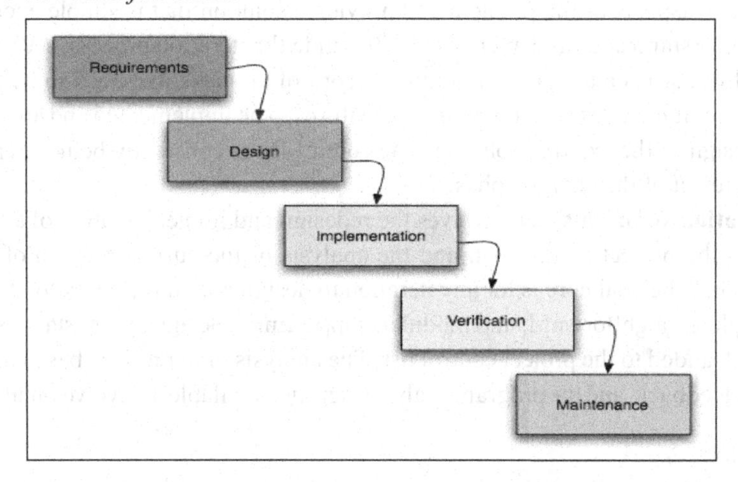

In the waterfall model, software development proceeds from one phase to the next in a purely sequential manner. The development process does not move to the next stage unless the previous stage has been completely signed off as absolutely completed. Thus the waterfall model maintains that one should move to a phase only when its preceding phase is completed and perfected. Phases of development in the waterfall model are thus discrete, and there is no jumping back and forth or overlap between them. However, there are various modified waterfall models (including Royce's final model) that may include several degrees of minor or major variations to the process.

Iterative Model

The Iterative model is an incremental software development process that was developed in response to the weaknesses of the more traditional waterfall model (see Figure 4.3). The two frequently used iterative development frameworks in this model are the Rational Unified Process and the Dynamic Systems Development Method.

The main idea behind iterative development is to develop a software system incrementally, allowing the developer to take advantage of lessons learned during the development of earlier versions of the system for later versions. Key steps in the process are to start with an implementation of a part of the software requirements and iteratively enhance the functionality and requirements until the full system is implemented.

Figure 4.4 shows the iterative model consisting of the following steps:

- **Initialization step:** This step creates a base version of the system. The goal is to create a product to which the user can react. It should offer a sample of the key aspects of the problem and provide a solution that is simple enough to understand and implement easily. To guide the iteration process, a VLITP control list is created that contains a record of all tasks that need to be performed. It includes such items as new features to be implemented and areas of redesign of the existing solution. The control list is constantly being revised as a result of the analysis phase.
- **Iteration step:** This step involves the redesign and implementation of a task from the project control list, and the analysis of the current version of the system. The goal here is for any iteration to design and implementation to be simple, straightforward, and modular, supporting redesign at that stage or as a task added to the project control list. The analysis of iteration is based upon user feedback, and the program analysis facilities available. It involves analysis

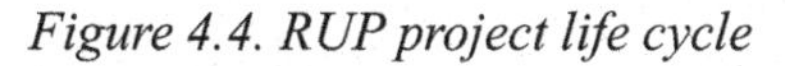

Figure 4.4. RUP project life cycle

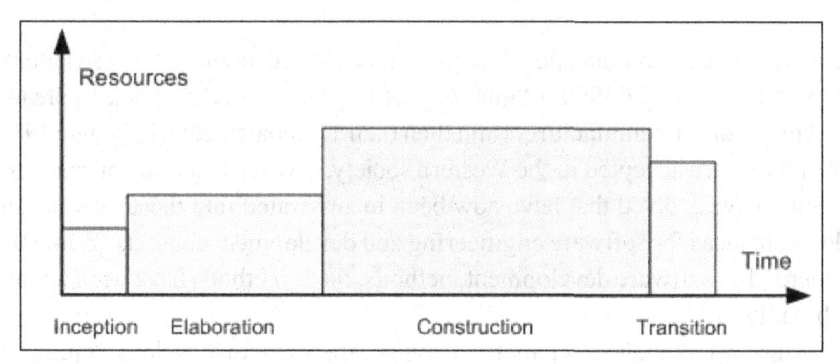

of the structure, modularity, usability, reliability, efficiency, and achievement of goals.

- **Project control list:** This stage is used to control any modification in light of the analysis results.

Guidelines that drive the implementation and analysis of this model includes:

- Any difficulty in design, coding and testing a modification should signal the need for redesign or re-coding.
- Modifications should fit easily into isolated and easy-to-find- modules. If they do not, some redesign is needed.
- Modifications to tables should be especially easy to make. If any table modification is not quickly and easily done, redesign is indicated.
- Modifications should become easier to make as the iterations progress. If they are not, there is a basic problem such as a design flaw or a proliferation of patches.
- Patches should normally be allowed to exist for only one or two iterations. Patches may be necessary to avoid redesigning during an implementation phase.
- The existing implementation should be analyzed frequently to determine how well it measures up to project goals.
- Program analysis facilities should be used whenever available to aid in the analysis of partial implementations
- User reaction should be solicited and analyzed for indications of deficiencies in the current implementation.

Agile Methods

The concepts of agile product development first emerged among Japanese automobile manufacturers in the 1980s (Aguanno, 2004). These concepts quickly spread to North American car manufacturers and then their IT departments in the late 1980s. As a result of there accepted in the Western society, several different approaches to the method have evolved that have now been incorporated into these new product development ideas for software engineering and development contexts. While they first emerged as software development methods, these methods have been adopted more broadly.

Agile project management methodologies are used to develop, deploy and acquire IT. It emerged into modern businesses much in the same way business process management did. This section applies agile methods in a VLITP as a high ceremony project management method. High ceremony projects are those based on formal or semi-formal project management methods (Aguanno, 2005). While some Agile development methods address the management aspects of software projects—people, processes and technology—it is primarily focused on coding, testing and software artefact delivery. Thus, applying the concept of agility to the management of software within VLITP environment is a natural step in the evolution of software development.

Agile methods allow VLITP manager, can to guide the sub-team leaders and continuously influence team behavior by defining, disseminating and sustaining a guiding vision. At the outset of the project, work closely with the customer to understand the vision for the project, how it is expected to support business goals, and how it will be used. To promote team ownership of the vision, facilitate a group discussion with the team to build a joint project vision. The strong grasp of VLITP vision helps the sub-project teams through difficult decisions about business value and priority and keeps them focused on and inspired by the ultimate goal.

Planning sessions are fertile ground for developing a common understanding and respect between the VLITP developers and the host organization. As the project progresses these sessions can become highly collaborative and creative resulting in improved morale and a better outcome of the VLITP. Basic facilitation techniques such as making sure members of various sub-teams have an opportunity to speak, summarize and confirm, and draw out concerns can help to build the each sub-project.

VLITP managers use agile method in leading sub-teams by establishing a guiding vision, fostering teamwork and cooperation, setting simple rules, championing open information, and managing with a light touch—with the understand that each team leader has his or her own ideas, and is likely to behave in accordance with those ideas. It requires the VLITP manager to be continually vigilant to merit the mantle of leadership: monitoring progress, and keeping a finger on the pulse of the development team. This does not mean hovering and controlling everything. Instead, it requires being observant, continuously seeking feedback and monitoring success or failure, and adapting by making changes as situations on the VLITP warrant:

- Reinforces the guiding vision of the VLITP at every opportunity by examining each sub-project decisions to see whether it lines up with the overall vision of the VLITP.
- Continually encouraging teamwork and collaboration between sub-project teams. Encouraging one-on-one discussions among team members as often as possible to keep things vibrant within the VLITP. It also helps to keep abreast with technology to facilitate the VLITP manager fully participating in conversations with software developers.
- Establishing simple rules while taking every opportunity to conduct process reflections: regularly examine what works and what needs improvement. Acting with courage to make changes when the VLITP manager feels the necessity.
- Working relentlessly to break down many barriers to information sharing that appear in VLITP implementation. It helps to keep apprised of cultural sensitivities, egos, and other such factors that may impinge upon a VLITP success. This may require operating with a light touch but being able to intervene quickly and wisely to solve personnel issues. Motivate and reward initiative, but manage expectations. Recognize and encourage self-organization, but disallow cliques.

Agile methods, at their core, focus on managing the impact of change on a project. When change occurs during VLITP, these methods provide ways of allowing that change to be introduced to the project in an orderly way that attempts to maximize the benefits for the sponsor, while controlling the risks that the change introduces.

In the other two product development lifecycles, design is developed before hand to facilitate the development of software product according to that design; then the software product is tested to determine how well it adhered to that design. Design changes introduced during the development or testing phases of a VLITP can often cause chaos. That's because they require the VLITP to stop dead in its tracks and occasionally cycle back to the beginning of the design phase. Imagine situations where such changes occur late in the VLITP, the resulting delays and increased costs can lead to impossible achievement of the VLITP objectives. Agile methods share the below common characteristics that allow them to respond better to change (Aguanno, 2005):

- **Iterative and incremental development.** Agile methods break down the development of new software products into several repeating cycles—iterations. Each iteration builds upon the previous one, adding additional layers of functionality. This allows the software to grow much in the same way that an corn grows by adding layers. Starting from an initial kernel of functionality, over time (and using several iterations) the software becomes more and more complete. Thus, allowing for the delivery of certain functionalities early on in the VLITP, and then progressively throughout various stages of the VLITP.
- **Short iterations.** By using short iterations, agile methods keep the feedback cycle short, allowing more responsiveness to change, and reducing the risk of building "the wrong thing" based on unclear or changing requirements.
- **Progress measured via completed features.** Rather than trying to track progress by measuring percent complete on intangible elements such as design, agile methods track progress by fully-completed and tested features. Progress is measured by the percent of features that have been complete and ready to be reviewed and subsequently deployed by the host organization. This has the advantage of delivery only software with full features that are fully and properly developed regardless of the timeline or budget constraint.
- **Open, flexible design.** Designs are flexible and extensible using open standards wherever possible. Since the full set of requirements may be unclear at the beginning of the project, and will likely change during the course of the VLITP, a flexible, extensible design is prepared that allows the addition of features to support new requirements which may subsequently emerge. There is always the risk of some rework to incorporate complex requirements to the VLITP, but often the impact is offset by other benefits gained when using agile methods.
- **Empowered teams.** Teams of specialists who know their jobs well and have the experience (and maturity) to decide for themselves how best to approach the problems at hand. Instead of imposing a design on the team, let the people who know best how to implement the details decide how they are going to do the work. This often allows for more flexibility in a system than is usually found in a more traditional, centralized design.
- **Personal communications.** Rather than focus on producing written documents to communicate design decisions, technical approaches, and other normally documented items, agile methods suggest that teams work in shared physical environments. Speaking face to face, perhaps with the support of a whiteboard for drawing diagrams, is a much more efficient way of working out design details. It is easier to update a design on a whiteboard, a design that is shared in the minds of a room full of people, than to update hundreds (or thousands) of pages of design documents. As we will see later in this chapter, many

large projects have failed because of the reliance on written documentation to communicate design. While written documentation has its place on agile projects teams need to seek opportunities to discuss approaches verbally to capture nuances not reflected always in documents, and to provide a fast way of asking (and answering) questions to ensure comprehension.

Using these common characteristics, the creators of the various agile methods have produced a variety of approaches, each emphasizing different aspects. No single agile method is "better" than the rest–they all have their strengths and weaknesses–yet employing them can often lead to some significant benefits for the sponsor and for the project team.

Benefits of Being Agile Methods

While agile methods may not fit every project, they do provide valuable benefits when applied against specific business problems. Consider each of these benefits when evaluating whether a particular agile technique may be of use on your own project, as each technique puts a different emphasis on the different benefits.

There are three main types of benefits from agile techniques: reducing risk, improving control, and improving communications. Each of these benefits has a number of aspects that are detailed below.

SOFTWARE DEVELOPMENT FRAMEWORKS

This section discusses the various software development frameworks used in VLITPs. It will focus on the following frameworks: RAD, RUP and TOGAF. There are several software development frameworks currently being used for implementing VLITPs. Three of these will be discussed in this section: RAD, RUP and TOGAF.

Rapid Application Development

Rapid Application Development (RAD) is a software development framework that focuses on building applications in a very short amount of time. Applications can be designed and developed within 60-90 days; RAD was originally intended to describe a process of development that involves application prototyping and iterative development.

The main disadvantages of using RAD compromise usability, features and/or execution speed. This is due to the fact that RAD uses time boxing. As a result, extra features are implemented to later versions and not part of initially delivered basic application in a short time frame. RAD may produce applications that do not contain the full feature as traditionally developed applications. VLITP managers should be concern this aspect of RAD.

Rational Unified Process

The Rational Unified Process (RUP) is an iterative software development process framework created by the Rational Software Corporation, to be adaptable by the project teams for the host organization. The sub-projects managers of a VLITP select the elements of the process that are appropriate for their needs. As software process product RUP includes a knowledge base with sample artifacts and detailed descriptions for many different types of business activities. The creators and developers of the process focus on diagnosing the characteristics of different critical software projects. That's because project failure is usually a result of a combination of several symptoms, though each project fails in a unique way. In consideration of this general belief, Rational Software Corporation undertook a study of best practices for system software and they named that Rational Unified Process.

RUP was designed with the same techniques used by project teams to design software with an underlying object-oriented model, using Unified Modeling Language (UML). RUP is based on the set of software development principles and best practices, listed below:

- Develop software iteratively
- Manage requirements
- Use component-based architecture
- Visually model software
- Verify software quality
- Control changes to software

The RUP lifecycle is available as a work breakdown structure, which could be customized to address the specific needs of sub-projects within a VLITP. The RUP lifecycle organizes the tasks into four phases and iterations as shown in Figure 4.4:

- Inception phase
- Elaboration phase
- Construction phase
- Transition phase

Inception Phase

In this phase the business case includes business context, success factors and financial report is made. To complement the business case, a basic use case model, project plan, initial risk assessment and project description are put on paper (Keil, 2000). After these are completed, the sub-project is checked with the help of the following criteria:

- Stakeholder concurrence on scope definition and cost/schedule estimates.
- Requirements understanding as evidenced by the fidelity of the primary use cases.
- Credibility of the cost/schedule estimates, priorities, risks, and development process.
- Depth and breadth of any architectural prototype that was developed.
- Actual expenditures versus planned expenditures.

If a sub-project does not pass the Lifecycle Objective Milestone, it would need to be reviewed or this phase will be redone after being redesigned to meet the criteria.

Elaboration Phase

The elaboration phase is where the sub-project really starts. In this phase the problem domain analysis is made and the architecture of the sub-project gets its form. This phase must pass the Lifecycle Architecture Milestone by meeting the following criteria:

- A use-case model in which the use-cases and the actors have been identified and most of the use-case descriptions are developed. The use-case model should be 80% complete.
- A description of the software architecture in a software system development process.
- An executable architecture that realizes significant use of cases.
- Business case and risk list which are revised.
- A development plan for the overall project.

Where a sub-project does not pass this milestone, there could be time for cancellation for redesign. After leaving this phase, the sub-project transcends into a high-risk operations where changes are much more complicated and may have a higher risk to the entire VLITP.

Construction Phase

In this phase the main focus goes to the development of various components and other features of the system being designed. This is the phase when the most of the coding takes place. Due to the nature of VLITPs, several iterations are normally developed for the purpose of smaller manageable segments that produce prototypes for demonstration. This phase produces the first external release of the software and concludes with the Initial Operational Capability Milestone.

Transition Phase

In the transition phase, the VLITP moves from the project management team to the host organization or end users. The activities within this phase include training of the end users, beta testing of the system to validate it against the end users' expectations (Deephouse et al, 1995). The software is also checked against the quality level set in the Inception phase. If it does not meet the standards of the end users, the entire cycle in this phase can be repeated. All objectives are must be met for the project to reached the Product Release Milestone in the development cycle.

RUP Disciplines

RUP is based on a set of building blocks which describe the following: what is going to be produced, the necessary skills that are required and a step-by-step explanation describing how specific development goals are going to be achieved. The main building blocks are the following:

- Roles (who)–A Role defines a set of related skills, competencies, and responsibilities.
- Work Products (what)–A Work Product represents something resulting from a task, including all the documents and models produced while working through the process.
- Tasks (how)–A Task describes a unit of work assigned to a Role that provides a meaningful result.

The iteration is where tasks are categorized into nine disciplines, six of which are defined as Engineering disciplines (Business modeling, Requirements, Analysis and design, Implementation, Test, Deployment) and three supporting disciplines (Configuration and change management discipline, Project management discipline, Environment discipline).

For each discipline there are several deliverables, milestones and procedures that need to be completed before a phase of the RUP project life cycle can be closed and the next phase can be started.

The Open Group Architecture Framework

TOGAF is a detailed method and a set of supporting tools used in VLITP for developing enterprise architecture. It is published and promoted by The Open Group and may be used freely by any organization wishing to develop an enterprise architecture. Because this documentation is a live TOGAF has different versions and is continuously being updated.

The main objective of TOGAF is to create Boundary Information Flow thus, enabling access to integrated information within and between enterprises based on open standards and global interoperability for software development. A number of disadvantages include the low speed at which it works, lack of real commitment, the inappropriate capacity of some the participating people, changes in organizational structures which makes it not very suited in today's fast changing world.

TOGAF framework consists of several building blocks including:

- Resource base
- The TOGAF foundation architecture consists out of the Technical reference, Standards Information Base and the Building Blocks Information Base.
- Architectural Development Method
- Target Architectures

VLITP MANAGEMENT TEMPLATE

Delivering an integrated approach for managing VLITPs involves several performance management and improvement issues. Some of these have listed below in a template for the VLITP with the views of ensuring assurance of meeting the strategic objectives.

Scope specification: The first, and most important, step in a VLITP is defining the scope of the project (Morris, 1998). What is it the host organization is supposed to accomplish by implementing this specific project? What are the major project objectives? Equally important is defining what is not included in the scope of the VLITP. If the host organization doesn't give a clear mandate to the project management team, clarification should be sought and confirmed before initiating the project (Deephouse et al, 1995).

Determination of available resources: The management team of a VLITP should determine what resources (including people, office space, equipment, money, etc) are available to facilitate the achievement of objectives (Rosenau and Githens, 2005). While the management team may not have direct and complete control over all these resources, an acceptable arrangement should be in place to manage resources without serious obstacle resulting from bureaucracy within the host organization. There should be a clear policy as to how these resources should be used for the VLITP.

Timeline indication: The timeframe of a VLITP should be defined and made cleared to all parties involved in the implementation process. When does the project have to be completed? As the project plan is being developed, there should be certain degree of flexibility in how the management team use time during the implementation, but deadlines usually are fixed (Heeks et al, 1999). In cases where the management team decides to use overtime hours to meet the schedule, it should be weighed against the limitations of total project budget.

Project team compilation: Assembling very good team of high quality IT professionals in the 21st century is a very big job. This can only be done if the project champion is successful in attracting a top IT professional to head the VLITP. That person should have certain degree of knowledge and influence in the IT industry. This will facilitate the initiation of dialogs with the wider IT community for suitable people to work on the sub-projects. They should include technical experts that would serve as functional supervisors assigned to various sub-projects. The job of properly managing VLITPs is usually half completed by assembling the right team.

Specification of major objectives: What are the major objectives of each sub-project? The management team should be cleared about various intervals for different sub-projects to be achieved. These should be cleared to all stakeholders to the VLITP and listed at strategic location in the project implementation sites, in chronological order (Rai, Lang and Welker, 2002). All involved with the VLITP should be aware when changes have been made to the order at a later date (Deephouse et al, 1995).

Indicate minor sub-projects: These are the smaller objectives that help to indicate if the major sub-projects have been achieved. Again, it usually helps the project management team to remember the entire set of minor sub-projects if they are listed in chronological order at various locations in on project sites. The level of details indicated more or less depends on the size and complexity of the project.

Development of preliminary plan: This requires the assembling of all the steps into a plan. Such plan would help to answer the following questions: What happens first in each sub-project? What is the next step? Which steps would be simultaneous using the same or different resources? Which sub-project is going to be responsible for each step? How long will each stage of a sub-project take? There are many good software packages available that can automate a lot of this detail for various sub-projects. The plan should benefit from previous VLITPs or members of the project management team with experience from similar positions.

Baseline plan approval: After getting feedback on the preliminary plan from all relevant parties and stakeholders to the VLITP, adjust the timelines and work schedules to fit the sub-project into the available time. Make any necessary adjustments to the preliminary plan that should result to the production of a baseline plan (Deephouse et al, 1995).

Documentation of project adjustments: There can never be sufficient time, money or talent assigned to a specific VLITP, requiring the management team to do more with the limited resources than expect (Rosenau and Githens, 2005). However, there are often limits placed on a VLITP that are simply unrealistic. The management team therefore needs to make the case for more resources and justify where unrealistic limits can be changed (Morris, 1998). Such changes should be requested at the beginning of the project and not when the project is in trouble to ask for the changes it needs.

Regular progress monitoring: It is important to monitor little progress at the beginning of every sub-project, and continue monitoring the activities of everyone on the team. This makes it easier to catch issues before they become problems (Kim, 2003; Rai, Lang and Welker, 2002). This is easier when VLITP incorporates in its plans a regularly monitoring policy.

VLITP implementation process documentation: Keep records of the various processes of the sub-projects as part of a VLITP. Every time the project faces change from the baseline plan, write down what the changes were and why they were necessary (Taylor and Todd, 1995). Every time a new requirement is added to parts of the sub-projects, it should be well documented. That would include the origins of the changes to project requirement and how the timeline or budget would be adjusted due to such changes. This is necessary because the management team can't obviously remember everything.

Soliciting everyone involvement: It is very important to keep all the stakeholders of VLITP informed of progress along the implementation process. By using Web sites that are updated regularly, the host organization can be kept aware of the successes at various aspects of the VLITP, during the completion of each milestone in sub-projects (Alpert, 2003). These avenues can also be used to inform various stakeholders of problems as soon as they occur. A good management team would inform the relevant parties when changes are being considered, as far ahead as possible. The point here is to ensure everyone on the team is aware of what everyone else is doing.

CHAPTER SUMMARY

The primary purpose of tools in VLITP is to avoid projects reaching a level of complexity where pragmatic management generates a level of inefficiency and waste that may threaten the final outcome of the project. Software development models and framework are technical disciplines required by all managers of VLITPs to run them as efficiently as possible. These different software tools can be applied along with more advanced project management methodologies to support the management of VLITPs.

This chapter has challenged such clichés by reconceptualizing the relation of VLITPs, rationality, and modernization of project management methodologies and framework. Its theme is the possibility of a truly radical reform of implementing VLITP in today's business environment.

The chapter has also argued that the degradation of labor, education, and the business environment is rooted not in implementing VLITPs per se but in the anti-modernization values that govern project management. Desirable as alternative goals may be, no fundamental progress can occur in the kind of the 21st century business environment that sacrifices millions of individuals to production. Essence of VLITP is to promote the value of individuals in a well-run organization taking advantage of all the modern technologies available to the host organization.

REFERENCES

Alpert, S. (2003). Protecting medical privacy: Challenges in the age of genetic information. *Journal of Social Issues, 59*(2), 301-322.

Aguanno, K. (2004). Ever-Changing Requirements: Use Agile Methods to Reduce Project Risk. *Proceedings Informatics 2004*. Canadian Information Processing Society (CIPS). May.

Charvat, J. (2003). Project Management Methodologies: Selecting, Implementing, and Supporting Methodologies and Processes for Projects, London: John Wiley and Sons.

Cleland, D. I. (1990). Project Management: Strategic Design and Implementation, TAB Books, PA, p. 23.

Deephouse, C., Mukhopadhyay, T., Goldenson, D. R., & Kellner, M.I. (1995). Software Processes and Project Performance. *Journal of Management Information Systems*,12(3), 187-205.

Heeks, R., Mundy, D., & Salazar, A. (1999). Why Healthcare Information Systems Succeed or Fail. *Information Systems for Public Sector Management Working Paper Series*, Paper Number 9.

Keil, M., Wallace, L., Turk, D., Dixon-Randall, G., & Nulden, U. (2000). An investigation of risk perception and risk propensity on the decision to continue a software development project. *Journal of Systems and Software, 53*, 145-157.

Kim, Y. (2003). Do Council-manager and Mayor-council Types of City Governments Manage Information Systems (IS) Differently? An Empirical Test. *International Journal of Public Administration, 26*(2), 119-134.

Morris, P. W. G. (1998). Key Issues in Project Management. In *Project Management Handbook* edited by J. K. Pinto, Jossey-Bass, p. 5.

Office of Government Commerce (2003). *Managing Successful Programmes*, published by HM Stationery Office, UK, (second edition), Foreword, p v.

Patel, M. B., & Morris, P. G. W. (1999). *Guide to the Project Management Body of Knowledge*, Center for Research in the Management of Projects, University of Manchester, UK, p-52.

Rai, A., Lang, S. S., & Welker, R. B. (2002). Assessing The Validity of IS Success Models: An Empirical Test and Theoretical Analysis. *Information Systems Research,* 13(1), 50-69.

Rosenau M., Githens G. (2005). *Successful project management: a step-by-step approach with practical examples*, Wiley, Hoboken, NY

Tao, L. (2000). Application Service Provider Model: Perspectives and Challenges. *Paper presented at the International Conference for Advances in Infrastructure For Electronic Business, Science and Education on the Internet (SSGRR)*, L'Aguila, Italy, July 31-August 6.

Taylor, S., & Todd, P. (1995). Decomposition and crossover effects in the theory of planned behavior: A study of consumer adoption intentions. *International Journal of Research in Marketing. 12*, 137-155.

Waterfall Model: http://en.wikipedia.org/wiki/Iterative_model (Accessed May 2007).

Wideman R. (2003). *Comparing PRINCE2 with PMBOK*, http://www.maxwideman.com/papers/comparing/intro.htm (Accessed May 2007).

Section II
Technology vs. People in VLITP

Chapter V
IT Governance

ABSTRACT

VLITP managers face unprecedented expectations for their governance. These expectations are driven by mandates and other demands from host organizations. This chapter is meant to help VLITP professionals around the world meet and exceed such expectations. It details relevant expertise, methodologies, and experience required by VLITP managers to go beyond compliance of regular IT governance issues to deliver objectives that drives business value across the host organization's enterprise. Implementing a VLITP involves the management of a transitional period which requires a structured approach that will help the host organization evaluate its options for designing the organizational structure that facilitates continuous business improvement. Good IT governance in VLITP focuses on immediate priorities, including a periodic identification and learning lessons to determine both near-term and far reaching strategies for the VLITP. This involves good approach to different stands and compliance issues during the implementation of VLITP.

INTRODUCTION

Organizations are faced with many risks such as natural disasters, information loss, fraud, and human errors (Brown, 2001). All these risks could cause an organization irreparable damage. To reduce the damage or even prevent the risk from happening, systems controls are necessary; thus making internal control a crucial element in the smooth operations of businesses. This frequently develops an increased need for internal control frameworks (i.e. COSO framework). The COSO framework is the most widely accepted and used framework for internal control. It is seen as the foundation for internal control within organizations based on internal control in general. When a VLITP becomes an important asset for the host organization, manual controls automatically shifts to IT controls, with needs for specific control framework. The last two decades have seen different IT control frameworks developed (such as CobiT, ITIL, ISO17799, and ISO20000). The chapter looks at CobiT is a general control framework for VLITPs, considering it is the most used IT control framework worldwide—mainly because it encompasses all aspects of IT. ITIL and ISO20000 are standards for IT Service Management in particular, while ISO17799 is a standard focused on Information Security, with CobiT.

The chapter also discusses the effective use of SOX Act in VLITPs. As host organizations work on complying with Section 404 of Sarbanes-Oxley Act of 2002 (SOX), which requires officers of a public company to establish, monitor and report on the effectiveness of the controls that ensure the integrity and accuracy of financial data, the challenge they face is not what IT controls to enforce and monitor. Instead, they deal with: identifying the scope of the network that is affected by this regulation; recognizing and standardizing on a set of controls that need to be applied uniformly across all affected IT systems; managing the volume of information that is required to demonstrate that compliance requirements have been met across the entire organization, in a changing threat environment (Trites, 2004). VLITP managers must continuously identify and profile all systems on the enterprise network to ensure all assets are managed and monitored for compliance. The process discloses all hosts, applications, services, and related vulnerabilities, providing a comprehensive view of the network and building the foundation for effective compliance management. It also enables the host organizations to set "Asset Values" to provide business context to the vulnerability scores and denote the value of a particular system in the organization. A system on the network, which is subject to SOX regulation, may have multiple vulnerabilities - each of which is assigned a vulnerability score. Managers of VLITP are able to use the same methodology for accurate, objective measurement of compliance across all networks.

INTERNAL CONTROL

Internal controls serves to verify whether information needed to make critical decisions in the organization are accurate and reliable. An organization's survival in a competitive economy greatly depends on the decisions made by management at all levels. These decisions are based on information they receive from various sources (i.e. financial reports). Organizations need to have the proper controls in place to facilitate their business processes functioning according to anticipation. Since most business processes are automated, also having automated controls systems has becomes inevitable. Thus, an IT control framework should be an effective tool for the identification and management of the necessary controls required in VLITPs.

Moreover, effective internal control can help the host organization to monitor business processes, assets and profitability. It can also deter negative behavior and reduce or prevent damage to the organization. Effective internal control can increase the credibility with shareholders and the public, especially that SOX Act has become the de-facto internal control requirement by US law and therefore cannot be absent in an organization, after numerous fraud and scandals in recent years.

The Committee of Sponsoring Organizations of the Treadway Commission (COSO) issued its final report on internal control in 1992 based on a three-year study. The COSO report (available at www.coso.org) contains the most widely accepted definition on internal control. Its definitions of internal control includes a process effected by an entity's board of directors, management, and other personnel designed to provide reasonable assurance regarding the achievement of objectives in the following three categories:

1. **Effectiveness and efficiency of operations:** Relating to the objectives of an organization including performance, profitability, and safeguarding of resources;
2. **Reliability of financial reporting:** addresses the preparation of reliable published financial statements under which interim and condensed financial statements and selected financial data derived from such statements, such as earnings releases, reported publicly
3. **Compliance with laws and regulations:** relating to laws and regulations to which an organization must comply.

Let us go through a couple of important points about the above definition of internal control:

1. **Internal control is a continuous built in component of operations.** Internal control is an ongoing series of activities that occur throughout an organization's operations, regardless of the fact that a VLITP is being implemented. As a recognized form of management controls, internal controls should be built into an organization as part of its infrastructure to help managers to run the organization and achieve their aims at an ongoing basis.
2. **People within an organization affect internal control.** People are one important component to build an organization. Therefore they are all responsible to some degree for internal control.

3. **Internal control can be expected to provide only reasonable assurance, not absolute assurance, regarding the achievement of the host organization's objectives for the VLITP.** No matter how well designed internal control is, it can only give reasonable assurance. There are internal and external factors that cannot be controlled such as human errors, collusion among employees, and external events beyond an organization's control.

The most widely known and accepted standard for internal control is the COSO internal control framework (Figure 5.1). The framework has grown to be considered as the foundation for internal control systems of may organizations.

The COSO framework consists of three dimensions as shown in Figure 5.1 as:

- The objectives.
- The components.
- The organization (units and activities).

To achieve the internal control objectives, an organization needs to have all the control components integrated in the organization's units and activities. In this context all three dimensions are related to each other. In order to realize these objectives the five control components of the COSO framework are necessary. These are:

1. Control environment.
2. Risk assessment.
3. Control activities.
4. Information and communication.
5. Monitoring.

Figure 5.1. Internal control framework

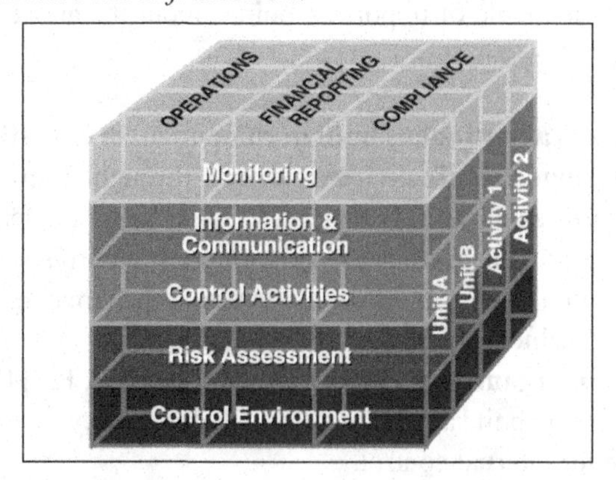

These control components are defined by the COSO report and are widely accepted as the minimum standards to which an organization should evaluate internal control at entity level. The next few pages will elaborate on each of these components will be addressed.

Control Environment

The control environment is the surroundings in which people operate in an organization. It is the foundation for all other components. That is why it is important for top management to create the right atmosphere. Only then there will be an effective control environment in which people understand that bad behavior is unaccepted and have consequences. Creating such an environment does not only require policies, procedures and code of ethics, but also management need to have a positive attitude and need to behave in an ethical way. Management needs to be an example for their staff.

Risk Assessment

A precondition to risk assessment is determining an organization's objectives. After setting the right objectives, risks associated with the achievement of these objectives should be identified (see Table 5.1). After identification, risk analysis should be performed to determine the possible effect these risks have and how they can be managed. Because conditions keep changing both inside and outside the host organization, a risk assessment needs to be performed on a continuous basis (see Table 5.1).

Control Activities

Control activities refer to the policies, procedures and mechanisms that are carried out to mitigate and manage risks (Rastogi and von Solms, 2005). These controls

Table 5.1. Overview risks in general projects and IT-projects

General risks in IT projects	Specific risks VLITPs
• Inadequate information • External events • Unclear/wrong goals • Inadequate resources • Failure of communication with management • Stress • Legal disputes • Scale	• Wrong requirements concerning IT • Political pressure • Risk management • Change and scope creep • Legacy issues • Unproven technology

can be either preventive or detective. In addition to that, they have different objectives and are implemented at different levels of an organization. They include common controls such as authorizations, security of assets, segregation of duties, and controls over information systems. The latter two are important for this thesis and will be further addressed.

1. Segregation of duties is necessary to reduce the risk of error or fraud. In this context, responsibilities for approving transactions, recording transactions, and handling the related asset should be separated from each other. All these tasks should be assigned to different people.
2. Controls over information systems can be divided into two groups. These are:
    - **General controls:** General controls relate to all information systems, including the applications that run on these systems. Information system can be mainframes, minicomputers, personal computer, networks, and telephone systems. General controls can be access security, change management, data center operations, and disaster recovery.
    - **Application controls:** Application controls relate to the processing of data within the application software and include input controls, processing controls and output controls. Input controls make sure that the recording of authorized transactions is complete and accurate. Processing controls make sure that processing of authorized transactions is complete and accurate. Output controls ensure reporting of complete and accurate audit trails of the results of processing.

Without effective general controls, application controls cannot be relied upon. For a complete and accurate data processing, an effective utilization of both of these controls is necessary in an organization.

SOX AND VLITPs

The need to handle data from untrusted sources with great care has added another layer of complexity to programmers job of writing secure software. VLITP managers must ensure there are tools to help programmers with this error-prone task. The normal approach is to extend the underlying type system to document the programmer's assumptions and to check that these are consistent with security. Similar challenge in writing secure software is managing one's privileges. Security-critical programs often must juggle various forms of privileges with unexpected associated pitfalls. There is a need for techniques for helping programmers with this task and for automatically deriving models of the semantics of operating system API's for managing privileges.

After a series of accounting scandals such (i.e. ENRON and WorldCom) the US congress adopted Sarbanes-Oxley Act (SOX) considered mandatory for all publicly traded companies in the US. The objective of SOX is to protect investors by improving the accuracy and reliability of corporate disclosures made pursuant to the securities laws, and for other purposes. More specifically, Section 404 of SOX requires management to take responsibility for establishing and maintaining an adequate internal control structure and procedures for financial reporting. Although SOX Section 404 makes no explicit mention of business continuity requirement, internal controls focusing on the availability and completeness of the financial reporting process are specifically addressed (Benvenuto and Zawada, 2004). So, it is the responsibility of management to make the necessary decisions regarding IT risks to insure the business continuity in order to produce timely, accurate, and complete financial reports.

VLITP managers must define a standard system configuration based on the operating system, applications and vulnerabilities of a host that complies with control objectives. All similar systems would then need to be continuously measured against such standard to ensure compliance. Deviations from the indicated standard due to the presence of an unauthorized application or outdated operating system can be efficiently identified and addressed based on the system's score determined by its vulnerability score and asset value (Feldman, 2004). This allows the host organization to intelligently prioritize resources to address the following critical areas of non-compliance:

- **Tracking changes:** Monitoring host configuration compliance during VLITP helps create an audit trail which may be used to demonstrate effective use of IT resources as well as levels of compliance to internal and external users. Tracking changes should be simple and preferably through the integration with relevant technologies which enable further integration of compliance management into existing enterprise processes (Brown and Grant, 2005).
- **Enterprise-wide compliance performance:** Managers of VLITPs often establish a robust reporting and risk analytics strategy for business executives as well as detailed operational reports that contain actionable intelligence for IT staff. The host organization can set compliance performance tolerances and measure against metrics such as vulnerability compliance, application compliance, remediation compliance, and coverage compliance. VLITPs in an open-architect system enable collecting and sharing security intelligence about the customer network with other information systems. Sharing intelligence

across the security ecosystem optimizes existing IT investments and enables organizations to better measure, manage and reduce non-compliance.

- **SOX compliance:** One of the greatest ongoing challenges VLITP mangers face when dealing with SOX is cost-effective resource management while reaching for continuous compliance. Several vendors offer solutions promising to help enterprises intelligently prioritize resource utilization by streamlining and automating administrative processes in data gathering and reporting (Brown and Grant, 2005). The host organization can set up pre-scheduled periodic scans of the network using tools built into these vendor software to automatically gather data and generate results to help assess the compliance posture of the network. Sub-project team managers are able to focus on measuring and managing network compliance instead of spending time on routine system administration, thereby becoming a strategic compliance resource and demonstrating higher return on compliance investment.

IT CONTROL AND SOX

SOX requires the chief executive and chief financial officers of public companies to attest to the accuracy of financial reports (Section 302) and require public companies to establish adequate internal controls over financial reporting (Section 404). Passage of SOX resulted in an increased focus on IT controls, as these support financial processing and therefore fall into the scope of management's assessment of internal control under Section 404 of SOX.

The 2007 SOX guidance from the PCAOB and SEC state that IT controls should only be part of the SOX 404 assessment to the extent that specific financial risks are addressed, which significantly reduces the scope of IT controls required in the assessment. This scoping decision is part of the entity's SOX 404 top-down risk assessment. In addition, Statements on Auditing Standards No. 109 (SAS109) discusses the IT risks and control objectives pertinent to a financial audit and is referenced by the SOX guidance.

The COBIT framework may be used to assist with SOX compliance, although COBIT IT controls that typically fall under the scope of a SOX 404 assessment may include:

- Specific application (transaction processing) control procedures that directly mitigate identified financial reporting risks. There are typically a few such controls within major applications in each financial process, such as accounts payable, payroll, general ledger, etc. The focus is on "key" controls (those that specifically address risks), not on the entire application.

- IT general controls that support the assertions that programs function as intended and that key financial reports are reliable, primarily change control and security controls;
- IT operations controls, which ensure that problems with processing are identified and corrected.

Specific activities that may occur to support the assessment of the key controls above include:

- Understanding the organization's internal control program and its financial reporting processes.
- Identifying the IT systems involved in the initiation, authorization, processing, summarization and reporting of financial data.
- Identifying the key controls that address specific financial risks.
- Designing and implementing controls designed to mitigate the identified risks and monitoring them for continued effectiveness.
- Documenting and testing IT controls.
- Ensuring that IT controls are updated and changed, as necessary, to correspond with changes in internal control or financial reporting processes.
- Monitoring IT controls for effective operation over time.

To comply with SOX, host organizations must understand how the financial reporting process works and must be able to identify the areas where VLITPs play critical parts. In considering which controls to include in the program, host organizations should recognize that IT controls can have a direct or indirect impact on the financial reporting process. For instance, VLITP application controls that ensure completeness of transactions can be directly related to financial assertions. Access controls, on the other hand, exist within these applications or within their supporting systems, such as databases, networks and operating systems, are equally important, but do not directly align to a financial assertion. Application controls are generally aligned with a business process that gives rise to financial reports. While there are many IT systems operating within an organization, SOX compliance only focuses on those that are associated with a significant account or related business process and mitigate specific material financial risks (see Table 5.2). This focus on risk enables management to significantly reduce the scope of VLITP general control testing in 2007 relative to prior years.

Section 409 requires public companies to disclose information about material changes in their financial condition or operations on a rapid basis. Public companies that are therefore implementing VLITPs need to determine whether their existing financial systems, such as enterprise resource management applications

Table 5.2. Real-time disclosure

Section	Title	Description
302	Corporate Responsibility for Financial Reports	Certifies that financial statement accuracy and operational activities have been documented and provided to the CEO and CFO for certification
404	Management Assessment of Internal Controls	Operational processes are documented and practiced demonstrating the origins of data within the balance sheet
409	Real-time Issuer Disclosures	Public companies must disclose changes in their financial condition or operations in real time to protect investors from delayed reporting of material events
802	Criminal Penalties for Altering Documents	Requires public companies and their public accounting firms to retain records, including electronic records that impact the company's assets or performance. Fines and imprisonment for those who knowingly and willfully violate this section with respect to (1) destruction, alteration, or falsification of records in federal investigations and bankruptcy and (2) destruction of corporate audit records.

are capable of providing data in real time, or if the organization will need to add such capabilities or use specialty software to access the data. The host organization must also account for changes that occur externally, such as changes by customers or business partners that could materially impact its own financial positioning (Cobb et al, 2007).

To comply with Section 409, host organizations should assess their technological capabilities in the following categories:

- **Availability of internal and external portals:** Portals help route and identify reporting issues and requirements to investors and other relevant parties. These capabilities address the need for rapid disclosure.
- **Breadth and adequacy of financial triggers and alert:** The organization sets the trip wires that will kick off a Section 409 disclosure event.
- **Adequacy of document repositories:** Repositories play a critical role for event monitoring to assess disclosure needs and provide mechanism to audit disclosure adequacy.
- **Capacity to be an early adopter of Extensible Business Reporting Language (XBRL):** XBRL will be a key tool to integrate and interface transactional systems, reporting and analytical tools, portals and repositories.

Section 802 and Records Retention

Section 802 of SOX requires public companies and their public accounting firms to maintain all audits or review work papers for a period of five years from the end of the fiscal period in which the audit or review was concluded. This includes electronic records which are created, sent, or received in connection with an audit or review (Brown and Grant, 2005). As external auditors rely to a certain extent on the work of internal audit, it would imply that internal audit records must also comply with Section 802.

In conjunction with document retention, another issue is that of the security of storage media and how well electronic documents are protected for both current and future use. The five-year record retention requirement means that current technology must be able to support what was stored five years ago. Due to rapid changes in technology, some of 20th century media still in use might be outdated in the next decade. Audit data retained using these media may not be retrievable not due to data degradation, but due to obsolete equipment and storage media.

Section 802 expects host organizations to respond to questions on the management of SOX content of VLITPs. IT-related issues include policy and standards on record retention, protection and destruction, online storage, audit trails, integration with an enterprise repository, market technology, SOX software and more. In addition, host organizations should be prepared to defend the quality of their records management program (RM); comprehensiveness of RM (i.e. paper, electronic, transactional communications, which includes emails, instant messages, and spreadsheets that are used to analyze financial results) , adequacy of retention life cycle, immutability of RM practices, audit trails and the accessibility and control of RM content during the implementation of VLITPs.

End-User Application / Spreadsheet Controls

PC-based spreadsheets or databases are often used to provide critical data or calculations related to financial risk areas within the scope of a SOX 404 assessment. Financial spreadsheets are often categorized as end-user computing tools that have historically been absent traditional IT controls. They can support complex calculations and provide significant flexibility. However, with flexibility and power comes the risk of errors, an increased potential for fraud, and misuse for critical spreadsheets not following the software development lifecycle (e.g. design, develop, test, validate, deploy). To remediate and control spreadsheets, public organizations may implement controls such as:

- Inventory and risk-rank spreadsheets that are related to critical financial risks identified as in-scope for SOX 404 assessment. These typically relate to the key estimates and judgments of the enterprise, where sophisticated calculations and assumptions are involved. Spreadsheets used merely to download and upload are less of a concern.

- Perform a risk-based analysis to identify spreadsheet logic errors. Automated tools exist for this purpose.
- Ensure the spreadsheet calculations are functioning as intended (i.e., "baseline" them).
- Ensure changes to key calculations are properly approved.

Responsibility for control over spreadsheets is a shared responsibility with the business users and IT (Brown and Grant, 2005). The IT organization is typically concerned with providing a secure shared drive for storage of the spreadsheets and data backup. The business personnel are responsible for the remainder.

IT CONTROL STANDARD

Another control model, but specifically for VLITP, is the Control Objectives for Information and related Technology (CobiT). The Information Security Audit and Control Foundation published the first version of this model. This Foundation together with the Information Security Audit and Control Association constitutes the IT Governance Institute. CobiT is a widely accepted standard for IT Governance and is based on a set of good practices for IT Governance.

CobiT's focus is on IT control aspects. It provides a link between business objectives and IT processes, IT resources and information. In this context the CobiT framework consists of three dimensions that are represented in the CobiT Cube (Figure 5.2). These are:

- **Information criteria:** Quality requirements, security requirements, and fiduciary requirements.
- **IT resources:** Data, application systems, technology, facility, and people.
- **IT processes:** Planning and Organization, Acquisition and Implementation, Delivery and Support, and Monitoring.

To clarify dimensions 1 and 3, they will be further addressed in the next subparagraphs.

INFORMATION CRITERIA

Information, a valuable asset to business processes, need to be delivered according to certain criteria. These business requirements for information are:

Figure 5.2. CobiT cube

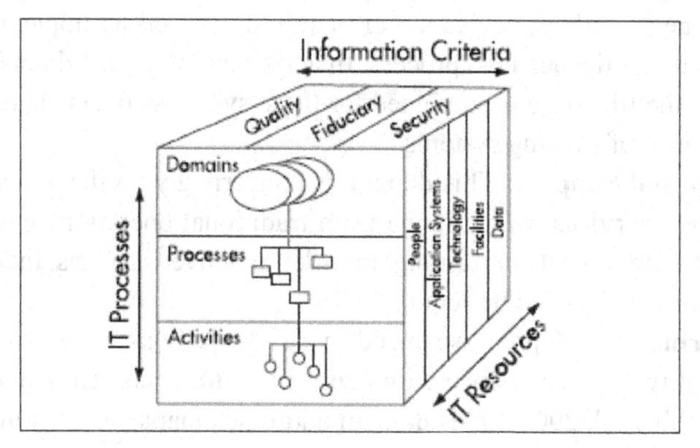

- Quality requirements
  - Effectiveness–Information needs to be relevant, correct, consistent and timely.
  - Efficiency–Information flow through the optimal use of resources needs to be efficient.
- Security requirements
  - Confidentiality–Information needs to be protected from unauthorized users.
  - Integrity–Information needs to be complete and accurate.
  - Availability–Information needs to be available when required.
- Fiduciary requirements
  - Compliance–Information needs to comply with laws and regulations.
  - Reliability–Information flow needs to be reliable.

IT PROCESSES

CobiT defines four domains in which the IT processes are grouped. These domains are:

- **Planning and organization:** This domain covers strategy and tactics, and concerns the identification of the way IT can best contribute to the achievement of the business objectives. Furthermore, the realization of the strategic vision needs to be planned, communicated and managed for different perspectives. Finally, a proper organization as well as technological infrastructure must be put in place.

- **Acquisition and implementation:** To realize the IT strategy, IT solutions need to be identified, developed or acquired, as well as implemented and integrated into the business process. In addition to that, this domain to make sure that the life cycle is continued for these systems covers changes in and maintenance of existing systems.
- **Delivery and Support:** This domain is concerned with the actual delivery of required services, which range from traditional operations over security and continuity aspects to training. In order to deliver services, the necessary support processes must be set up.
- **Monitoring:** All IT processes need to be regularly assessed over time for their quality and compliance with control requirements (Brown and Grant, 2005; Cobb et al, 2007). This domain addresses management's oversight of the organization's control process and independent assurance provided by internal and external audit or obtained from alternative sources.

CHAPTER SUMMARY

This chapter has looked at several levels of IT governance issues that affect VLITP. By this means project management can consider short-term planning because of the need to function properly on daily, weekly and monthly basis while the host organization is going through massive changes to comply with national and international regulations. While some these types of planning takes place the lower levels of project management and supervision with maximum emphasis on the practical aspects of VLITPs, the host organizations have to consider legal implications regarding SOX during VLITPs implementations.

The chapter also reviewed CobiT's focus is on IT control aspects by providing links between the overall business objectives for VLITP and IT processes, IT resources and information. This context is demonstrated in the CobiT framework consisting of a three-dimensional cube (Figure 5.2). On the other hand, IT controls require standardization and must follow criteria that fit with the necessary IT processes.

REFERENCES

Brown, A. E., & Grant, G. G. (2005). Framing the Frameworks: A Review of IT Governance Research. *The Communications of the Association for Information Systems*, *15*(37).

Brown, T. (2001, November). Modernization or failure? IT development projects in the UK public sector. *Financial Accountability and Management, 17*(4), 363-381.

Benvenuto, Nicholas and Brian Zawada. 2004. The Relationship Between Business Continuity and Sarbanes-Oxley. September, 2006. http://www.protiviti.com/knowledge/current_feature/031204.html

Broadbent, M. (2003, April). The Right Combination. *CIO Journal*, 13-14.

Cobb, A. T., Guan, J., & Levitan, A. S. (2007). Control Considerations in Object-oriented Systems. *Information Systems Control Journal, 3*, 18.

Feldman, M. S. (2004). Resources in Emerging Structures and Processes of Change. *Organization Science, 15*(3), 295-309.

Rastogi, R., & von Solms, R. (2005). Security Management, Integrity, and Internal Control in Information Systems. *IFIP International Federation for Information Processing*, Boston.

Trites, G. (2004). Director Responsibility for IT governance. *International Journal of Accounting Information Systems*, *5*, 89-99.

Weill, P., & Ross, J. (2004). Don't Just lead, Govern: How Top-Performing Firms Govern IT. *MIS Quarterly Executive, 3*(1).

Chapter VI
IT Security

ABSTRACT

One area that has scarcely received attention in the IT security literature, is the role that individual compliance plays in preventing cyber-attacks. Specifically, how individuals take precautions, how they are motivated to take precautions, and the impact of corporate security policies on individual precaution-taking behaviour have not been extensively researched. Existing literature has underdeveloped conceptualizations of how these control systems work in the realm of information security. This chapter adds to the body of knowledge concerning the socio-organizational perspective for understanding IT security management in the organization that implement VLITP. It examines the VLITP implementation process for achieving IT security management BS 7799 Part 2 certification. The author also gives regards to the role of individual perceptions of the compulsion of controls as a significant part of the IT security process. Focusing more on behavioural aspects of security during the implementation of VLITP, this book considers Information security is to be different from computer security—which is the encompassing of information security in addition to the other aspects of security such as technical aspects, physical security, system security, networking issues, and so forth.. IT security risk considerations cause are capable of causing particular concern on the interdependence of IT systems and inject another element of complexity in the application of the policies governing VLITPs.

INTRODUCTION

The primary purpose of this chapter is to identify the best approaches for integrating security into the delivery process of VLITP and improve the security of the facility throughout its life cycle. A secondary, though no less important, purpose is to provide a method to assess the impacts of common security approaches on key business outcomes for every sub-project objectives—including project cost, schedule, safety performance, etc.

Important concepts in the security objectives of VLITP deal with the issue of personal information—a valuable asset for doing business over the Internet. The host organization relies on its customers' personal information not only to enable basic transactions and operations of their business but also to identify new business opportunities. Many risks could be involved in its e-commerce transaction such as poor product quality, unauthorized sharing of personal information, among others. We focus, in this chapter, on privacy risks relating to IT security when implementing VLITP by looking at two privacy beliefs formed from the assessment of privacy risks:

- Privacy protection is the subjective probability that customers of the host organization believe that their private information is fully protected according to their expected high standard (Pavlou and Chellappa 2001).
- Privacy risk on the other is where the host organization recognizes a potential loss associated with releasing personal information to the service provided implementing VLITP (Malhotra et al, 2004).

The above contrary privacy beliefs reflect different aspects of VLITP security risk assessment and their separation may allow the examination of the data privacy issues more closely. While both privacy beliefs may seem related, they are often driven or wrought by different factors; thus, play different roles in influencing IT security policies, behaviors and decisions during the implementation of VLITP. Although privacy protection belief is not related to the explicit benefits of the primary exchange, consumers with a high privacy protection belief should perceive more control over privacy risks and are more likely to disclose customers' personal information. Conversely, host organizations in highly critical industry (i.e. healthcare or defense), quite often perceive a greater loss potential and may be wary about the disclosure of their customers' personal information during the implementation of VLITP.

BS 7799 stems from the publication of A Code of Practice for Information Security Management in 1993 and then of BS 7799 Part 1 in 1995 in the United Kingdom. It emphasizes more on the development of an IS security management framework and policy, than the technical requirements of IT projects. While previous success of BS 7799 (Part 1) has led to its transformation into an international standard ISO/IEC 17799 (in 2000), BS 7799 Part 2 remains the associated certificate scheme (developed in December 2005) as ISO/IEC 27001. Backhouse et al (2006) describes the institutionalization process of BS 7799 at industry and international levels. This chapter examines the human aspects and organizational issues of BS 7799 during the implementation of VLITP and achieving BS 7799 Part 2 certificate as an evidence of institutionalize IT security management practice in the host organization.

The increased attention to relevance of context in studying IT security phenomenon by IT security researchers today influences the need to make this an important part of VLITP plan, although such theoretical approaches are still at a theory-building stage (Dhillon & Backhouse, 2001). While many existing literature emphasize the external forces on managing IT security others have focused on the institutionalization process in the host organization. Hu et al (2006) examined the impact of institutional isomorphism on the adoption of IT security management practice. Siponen and Iivari (2006) argue that the stringent rule-based and context-free security policy is not applicable in today's turbulent and fast-changing business environment. They elaborate that to maintain competitive advantage, while coping with unforeseen business circumstances, employees of modern organizations might not have the patience to follow the formal compliance process for approval to information access. These authors suggest that the design of IT security policies requires the input of "application principles to solve such exceptional situations" (Siponen and Iivari, 2006: 448) instead of enforcing IT security policies literally. Willcocks and Margetts (1994) address the significance of contextualism when undertaking the risk assessment in the organization. Dhillon and Backhouse (2001) propose the analytical tool of responsibility structure to examine the norms and patterns of behaviors of different actors in the organization.

Within the last two decades IT security has received a great deal of attention in popular press. Straub and Welke (1998) have also documented the risks of having poor security during VLITP, either in the form of identity theft, data loss, or misappropriation of IT resources. Other studies have shown a marked increase in data theft and the creation of malicious code developed specifically to steal confidential information (see Symantec Corporation, 2007). VLITP cannot be exposed to cyber criminals who are continuing to refine their attack methods to remain undetected and to create global, cooperative networks to support the ongoing growth of criminal activity (Symantec Corporation, 2007). Many host organization view security as one of their top concerns. Despite the efforts of project mangers to secure systems and data, exposure to potential hackers has been observed in increasing numbers when IT security is not deemed a top priority or the nature of risk is not properly realized (Frieze et al. 1987; Swartz, 2005).

By Information security the author refers to all necessary measures that assure that IS will behave as expected and produce reliable results (Garfinkel et al. 2003; Ross 1999). Host organizations often address issues of IT security through technical means—by using centralized firewalls and other package software in an effort to protect business data. Achieving secure systems requires more than simply focusing on the technical issues. It also requires that top management gives attention to the design of effective IT security policies and motivate staff behavior in following such policies (Dutta & McCrohan 2002; National Cyber Security Alliance, 2005). While extensive measures to are being put in place to protect information systems, staff often bypass extant information security policies; thus, exposing the host organization to data loss and cyber crime (Dhillon, 2001). This introduces an additional problem of how to promote security policies amongst staff in the most effective way through the appropriate procedures.

For several decades both practitioners and theoreticians have had concerns that Information security has not received the required priority during VLITP (Dhillon 2001; Hughes & DeLone, 2007; Pearson & Weiner, 1985). Due to the current trend of implementing VLITP within interconnected networks to integrate all aspects of businesses while reaching outside organizational walls—via the Internet—organizations are exposed to potential attacks in new ways that are often misunderstood (Whitman, 2003). Such exposure resulting from Internet connectivity, host organizations are demand the implementation of wide array of technical safeguards such as firewall software and hardware and intrusion detection devices (Dhillon & Backhouse (2001); Dutta & McCrohan 2002).

These technical solutions are often the focus of organization security efforts simply because of the ease of implementation and centrally control. However, they are not always the most effective means of securing systems during a VLITP. While aspects of the system can be securely defended through technical means, technical safeguards could quite easily be breached when a single user shares a password or opens an e-mail containing a new virus that is not included in the anti-virus software.

On the other hand, behavioral solutions, such as educating individuals about security measures and informing others of penalties for disregarding security policies, have been found to be more effective than technical approaches but also more difficult to effectively implement (Straub, 1990). Other studies considering these behavioral solutions typically view it as secondary to technical means of preventing cyber attacks (Chin 1999; Ives et al. 2004; Mercuri 2002; Straub & Welke, 1998). Very few studies look directly at the role that behavioral policies and procedures play in the process of encouraging employees to secure assets (Hone & Eloff 2002; Kankanhalli et al. 2003; Kotulic & Clark 2004).

The ultimate goal of implementing security policies and procedures during a VLITP is for staff to follow those policies and take the necessary precautions. Without staff collaboration, security policies are meaningless with absolutely no influence on the overall security of the systems being implemented. The author's use of precaution here refers to the degree at which the host organization perceives certain measures to secure IT as well as dealing with information security in accordance with prescribed corporate security policies and procedures through specific proactive actions. Further to following the prescribed security policies and procedures, staff of the host organization must be aware of security threats at all times. Straub (1990) warns that such awareness can be enhanced through the formation and communication of formal information security policies.

SECURITY THREATS

As a result of numerous security incidents during the implementation of VLITP, host organizations today are demanding that VLITP implementation plans meet the requirements of regulatory compliance—such as Sarbanes-Oxley Act of 2002 or the Data Protection Act (PriceWaterhouseCoopers, 2006; Gordon et al., 2006). IT security literature shows a number of studies investigating issues regarding IT security management in organizations—focusing mostly on the impact of preventive and deterrent effort on security effectiveness (Straub, 1990; Kankanhalli et al., 2003), the design of IT security policy (Whitman et al., 2001; Siponen & Iivari, 2006), or the issue of management and employee awareness (Siponen, 2000; Straub & Welke, 1998). Dhillon and Backhouse (2001) classified security research paradigms by applying the Burrell and Morgan framework and revealing the dominance of technical and functionalist approaches in the existing security literature. They concluded that a socio-organizational perspective is the best way to achieve security of information systems (Dhillon and Backhouse, 2001). Thus, security literature in the functionalist paradigm still need improve on the social nature of the organizational problem of securing a system. The situation with personal data disclosure in the UK by the end of 2007 exemplifies the disastrous consequence of ignoring socio-organizational issues when implementing IT security management (Guah, 2008).

An SVA also employs the concepts of *threats* and *consequences* to assess security vulnerability. A threat is defined as any indication, circumstance, or event with the potential to cause loss of, or damage to, an asset. It also includes the intention and capability of an adversary to undertake actions that would be detrimental to valued assets. Adversaries might include: terrorists, either domestic or international;

activist or pressure groups; criminals (e.g., white-collar, cyber hackers, organized, opportunists). Sources of threats may include: insider, external, and insiders working as colluders with external sources.

Implicit in the threat concept is likelihood of the event occurring. As the threat increases, the likelihood of the security incident increases, as well. Threat ratings range from 1, very low, to 5, very high. Very high indicates that a definite risk exists and that the adversary has both the intent and capability to breach security possibly resulting in the consequences listed in Table 5. It also indicates that the facility, or similar assets, is targeted on a recurring basis. Very low, on the other hand, suggests no credible evidence of intent or capability, and no history of actual or planned threats against a facility or similar assets.

Importance of Cyber Security

Rampant federal government computer break-ins, denials of service, and general disruption of computer services has been causing increasing concern that these services can continue to operate and grow effectively unless further actions are taken to mitigate the risks. In partial answer to public concern about these issues in both the private and public sector, guidance and direction on Computer Security has been provided in the US President's January, 2000 statement (National Plan for Information Systems Protection). Among other general provisions, this plan

Table 6.1. Criteria for rating security threat to VLITP

Threat Level	Description	Consequences
Very High	Indicates that a ***definite*** threat exists against the asset and that the adversary has both the capability and intent to launch an attack or commit a criminal act, and that the subject or similar assets are targeted on a frequently recurring basis.	Very Severe
High	Indicates that a ***credible*** threat exists against the asset based on knowledge of the adversary's capability and intent to attack or commit a criminal act against the asset, based on related incidents having taken place at similar assets or in similar situations.	Severe
Medium	Indicates that there is a ***possible*** threat to the asset based on the adversary's desire to compromise similar assets and/or the possibility that the adversary could obtain the capability through a third party who has demonstrated the capability in related incidents.	Moderate
Low	Indicates that there is a ***low*** threat against the asset or similar assets and that few known adversaries would pose a threat to the asset	Minor
Very Low	Indicates ***no credible*** evidence of capability or intent and no history of actual or planned threats against the asset or similar assets	Very minor

outlines a number of new, centrally managed entities and projects that have been initiated to assist agencies in strengthening their security programs and improving federal intrusion detection capabilities. In addition, on March 3, 2000, in response to Internet disruptions, the US President also issued a memorandum to the heads of executive departments and agencies urging them to renew their efforts to safeguard their computer systems against denial-of-service attacks from the Internet.

SECURITY STANDARDS

Research work by Chapman and Leng (2004) has produced a three-step protocol for developing a risk mitigation plan to optimize protection of constructed facilities. Their protocol has been seen to help decision makers assess the risk of their facility to damages from natural and man-made hazards. This is done by identify engineering, management, and financial strategies for abating the risk of damages; and use standardized economic evaluation methods to select the most cost- effective combination of risk mitigation strategies to protect their facility.

Information security policies and practices provide the framework to protect an organization's computer-supported resources and assets. This protection ensures the integrity, appropriate confidentiality, and availability of the data and systems of an organization. Integrity ensures that data have not been altered or destroyed in an unauthorized manner. Confidentiality ensures that information is not made available or disclosed to unauthorized individuals or entities. Availability ensures that data will be accessible or usable upon demand by an authorized entity. Key activities for managing information security are risk assessment, awareness, controls, evaluation, and central management. Risk assessments consist of identifying threats and vulnerabilities to information assets and operational capabilities, ranking risk exposures, and identifying cost-effective controls. Awareness involves promoting knowledge of security risks and educating users about security policies, procedures and responsibilities. Evaluation involves monitoring effectiveness of controls and awareness activities through periodic evaluations. Central management involves coordinating security through a centralized group.

SECURITY REQUIREMENTS

A systems development environment is also defined so that each VLITP can apply the elements of that environment for consistent management and control. Systems development is the most sensitive part of VLITPs, and most failures occur because the development of software never meets expectations, or cannot be made operational. To provide the software needed for a VLITP, the organization can develop software using its own staff, use a contractor to develop software, use commercial off-the-shelf software, or a combination. To effectively manage software development and acquisition processes, however, the organization needs to have well defined software management processes, including methodologies and standards that will be used. Key processes for software development include requirements management, project planning, project tracking and oversight, quality assurance, and configuration management. For software acquisition, additional management is needed for solicitation, contract tracking and oversight, product evaluation, and transition to implementation and support.

Facility security, like many other facility attributes, can be enhanced most cost effectively when addressed early in the planning and design phases of a project. While this assertion may seem obvious, this study went far beyond confirming this contention and identified specific activities during project delivery that can be used to improve facility security and provide a quantitative assessment of the integration of security into established processes.

The validation process of a well-secured VLITP serves the following purposes:

1. To evaluate whether this tool will be an effective means of assessing security.
2. To quantify the impact of security best practices on cost, schedule, and safety.
3. To establish longer term trends in security integration within the host organization.

After several IT security incidents, GAO (government audit office in the US) highlighted six areas of management and general control problems in computer security:

- Poor security planning and management is the rule rather than the exception.
- Most agencies do not develop security plans for major systems based on risk, have not formally documented security policies, and have not implemented programs for testing and evaluating the effectiveness of controls they rely on.
- These are fundamental activities that allow an organization to manage its information security risks cost-effectively rather than by reacting to individual problems ad hoc.

- Agencies often lack effective access controls to their computer resources (data, equipment, and facilities) and, as a result, are unable to protect these assets against unauthorized modification, loss, and disclosure. These controls would normally include physical protections such as gates and guards and logical controls, which are controls built into software that (1) require users to authenticate themselves through passwords or other identifiers and (2) limit the files and other resources that an authenticated user can access and the actions that he or she can take.
- Application software development and change controls are weak. For example, testing procedures are undisciplined and do not ensure that implemented software operates as intended, and access to software program libraries is inadequately controlled.
- Agencies lack effective policies and procedures governing the segregation of duties. It is commonly found that computer programmers and operators are authorized to perform a wide variety of duties, such as independently writing, testing, and approving program changes. This, in turn, provides them with the ability to independently modify, circumvent, and disable system security features.
- Reviews frequently identify systems with insufficiently restricted access to the powerful programs and sensitive files associated with the computer system's operation, e.g., operating systems, system utilities, security software, and database management system. Such free access makes it possible for knowledgeable individuals to disable or circumvent controls.
- Service continuity controls are incomplete and often not fully tested for ensuring that critical operations can continue when unexpected events (such as a temporary power failure, accidental loss of files, major disaster such as a fire, or malicious disruptions) occur.

This was a typical example of the emerging cross cutting issue of international cyber security. It is not related directly to the formulation, development, and management of a single VLITP, but requires the focus of IT management at all levels to address the potential vulnerabilities across the entire world-wide electronic infrastructure. In that sense, it is very similar to the Y2K issue, and will have a large impact on the formulation and management of VLITPs during the next several years, as well as the retrofitting of the existing IT infrastructure. The lesson here is that we now must provide more focus on cyber security risk, vulnerabilities, and architecture in the design of VLITPs. If we don't change, we can predict more spectacular failures due to these vulnerabilities in the future.

CHAPTER SUMMARY

Security threat is not static throughout VLITP delivery cycle. It is linked to consequences of a security breach. As the project progresses or outcome indicators change, the threat to the VLITP may change as well. Security-related enhancements of VLITP are less costly and more effective when integrated early in the project life cycle. This allows the scope was further refined to include the following project phases—front end planning, detailed design, procurement, construction, and start-up. After several reported IT security incidents, GAO (government audit office in the US) highlighted six areas of management and general control problems in computer security

Facility security, like many other facility attributes, can be enhanced most cost effectively when addressed early in the planning and design phases of a VLITP. This is done by identify engineering, management, and financial strategies for abating the risk of damages; and use standardized economic evaluation methods to select the most cost- effective combination of risk mitigation strategies to protect their facility.

Many security risks could affect a VLITP outcome such as poor product quality, undelivered objective, unauthorized exposure of company data to competitors, complaints from customers, among others. This chapter has focused on privacy risks relating to IT security when implementing VLITP by looking at two privacy beliefs formed from the assessment of privacy risks:

REFERENCES

Backhouse, J., Hsu, C., & Silva, L. (2006). Circuits of Power in Creating De Jure Standards: Shaping an International Information Systems Security Standard. *MIS Quarterly, 30*(Special issue), 413-438.

Chin, S. K. (1999). High-confidence design for security. *Communications of the ACM 42*(7), 33-37.

Department of Trade and Industry, and PricewaterhouseCoopers (2006). *Information Security Breaches Survey 2006—Technical Report*, p. 36.

Dhillon, G. (2001). Violation of safeguards by trusted personnel and understanding related information security concerns. *Computers & Security, 20*(2), 165-172.

Dhillon, G., & Backhouse, J. (2001). Current Directions in IS Security Research: Towards Socio-Organizational Perspectives. *Information Systems Journal* (11), 127-153.

Dhillon, G., & Torkzadeh, G. (2006). Value-focused Assessment of Information System Security in Organizations. Information Systems Journal, (16), 293-314.

Dutta, A., & McCrohan, K. (2002). Management's role in information security in a cyber economy. *California Management Review, 45*(1), 67+.

Gordon, L., Loeb, M., Lucyshyn, W., & Richardson, R. (2006). *CSI/FBI Computer Crime and Security Survey Computer Security Institute*, p. 29.

Guah, M. W. (2008). Changing Healthcare Institutions with Large Information Technology Projects. *Journal of Information Technology Research*, 1(1), 14-26.

Hone, K., & Eloff, J. H. P. (2002). Information security policy - what do international information security standards say? *Computers & Security, 21*(5), 402-409.

Hu, Q., Hart, P., & Cooke, D. (2006). The Role of External Influences on Organizational Information Security Practices: An institutional perspective. *39th Hawaii International Conference on System Sciences,* Hawaii, pp. 1-10.

Hughes, L. A., & DeLone, G. J. (2007). Viruses, worms, and Trojan horses - Serious crimes, nuisance, or both? *Social Science Computer Review, 25*(1), 78-98.

Ives, B., Walsh, K. R., & Schneider, H. (2004). The domino effect of password reuse. *Communications of the ACM, 47*(4), 75-78.

Kankanhalli, A., Teo, H. H., & Wei, K. K. (2003). An Integrative Study of Information Systems Security Effectiveness. *International Journal of Information Management*, pp. 139-154.

Kotulic, A. G., & Clark, J. G. (2004). Why there aren't more information security research studies. *Information & Management, 41*(5), 597-607.

Malhotra, N. K., Kim, S. S., & Agarwal, J. (2004). Internet Users' Information Privacy Concerns (IUIPC): The Construct, the Scale, and a Causal Model. *Information Systems Research, 15*(4), 336-355.

Mercuri, R. T. (2002). Security watch—Computer security: Quality rather than quantity. *Communications of the ACM, 45*(10), 11-14.

National Cyber Security Alliance (2005). Top Ten Cybersecurity Tips. *National Cyber Security Alliance*, Washington DC.

Pavlou, P. A., & Chellappa, R. K. (2001). *The Role of Perceived Privacy and Perceived Security in the Development of Trust in Electronic Commerce Transactions.* Marshall School of Business, USC, Los Angeles.

Pearson, F. S., & Weiner, N. A. (1985). Toward an Integration of Criminological Theories. *Journal of Crime and Criminology, 76*(1), 116-150.

Siponen, M. (2000). A Conceptual Foundation for Organizational Information Security Awareness. *Information Management & Computer Security, 8*(1), 31-41.

Siponen, M., & Iivari, J. (2006). Six Design Theories for IS Security Policies and Guidelines. *Journal of Associations for Information Systems*, 445-472.

Straub, D. (1990). Effective IS Security: An Empirical Study. Information Systems Research, 1(3), 255-276.

Straub, D. W., & Welke, R. J. (1998). Coping with systems risk: Security planning models for management decision making. MIS Quarterly, 22(4), 441-469.

Symantec Corporation (2007). Symantec Reports Rise in Data Theft, Data Leakage, and Targeted Attacks Leading to Hackers. *Financial Gain.*

Whitman, M., Townsend, A., & Aalberts, R. (2001). Information Systems Security and the Need for Policy. In *Information Security Management: Global Challenges in the New Millennium,* G. Dhillon (ed.), IDEA Group Publishing, Hershey, 2001.

Whitman, M. E. (2003). Enemy at the gate: Threats to information security. *Communications of the ACM, 46*(8), 91-95.

Willcocks, L., & Margetts, H. (1994). Risk Assessment and Information Systems. European Journal of Information Systems, (3) 127-139.

Chapter VII
Human Resource Issues in VLITP

ABSTRACT

The study of diffusion, adoption, and IT project implementation in popular literature relies on theories which do not address the question of why VLITP projects continue to experience delays and go over budget simply due to problems with human resources. This chapter considers that such situation is only bound to continue unless specific research work are dedicated to investigating human resource management issues involved when implementing VLITPs. It details that VLITP are likely to miss the schedule date as a result of underestimating the length of time or the amount of resources required for various tasks, but omitting a task by an incompetent member of the VLITP team could lead to an incomplete or unaccomplished outcome. Avoiding mistakes in the sequencing of project tasks can ensure VLITP meet schedule. It suggests that VLITP managers build the project schedule by orderly listing major tasks and dedicating adequate human resources that considers several different kinds of eventualities.

INTRODUCTION

The management team of a VLITP must effectively manage the resources assigned to the project. This includes the labor hours of the systems designers, the code builders, the testers and the inspectors on the various sub-projects. Human resource (HR) management also includes managing issues related to the performance of associated subcontractors. Managing VLITPs resources frequently involve more than people management. This chapter details what VLITP managers must concentrate on regarding the needs of people—project employees, vendor staff, subcontract labor—such as equipment, vehicles, communication equipment, development and testing tools, staging servers, CD burners, disks and tape, manuals, laptops, mobiles and ipods, etc. decks, mixers, microphones and speakers. Managing the people resources means having the right people, with the right skills and the proper tools, in the right quantity at the right time (Cleland, 1990).

It is widely believed that VLITP is condemned to authoritarian management, mindless work, and equally mindless waste of money (Martin et al, 2005). Critics claim that VLITPs rationality and the level of manpower are contending for the soul of businesses in the 21st century. This chapter challenges such clichés by re-conceptualizing the relation of VLITPs, rationality, and modernization. Its theme is the possibility of a truly radical reform of doing business in today's environment.

It argues that the degradation of labor, education, and the business environment is rooted not in implementing VLITPs *per se* but in the anti-modernization values that govern project management. Reforms that ignore this fact usually fail, including such popular notions as a simple project for isolated functions in single unit of multi-national organizations. Desirable as these goals may be, no fundamental progress can occur in the 21st century that sacrifices millions of individuals to unrecognised/unappreciated contribution. The essence of human resource management in VLITP is to promote the value of individuals in a well-run project taking advantage of all the modern technologies available to them.

HUMAN RESOURCE MANAGEMENT

Providing satisfactory resources for VLITPs goes beyond a simplistic economic objective for the project. Considering VLITPs are managed by people with huge salaries the needs of many lower level staff that have interests in their own benefits as in benefits for the host organization must also be protected (Maslow, 1943). Consequently staff working on VLITPs frequently set objectives that fulfil their own interests as well as meeting objectives at the level that will be realistic to the all stakeholders (Forsberg, Mooz and Cotterman, 2000). As a result of these two separate interests, there are sometimes bargaining and internal politics in the formulation of project management objectives to reconcile the two objectives (Eisenhardt, 1996; Maslow, 1943). Managers of VLITPs consider the important fact that their personal objectives can only be achieved through employment when the host organization is happy that its objectives are being met to a reasonable level; though, it is not always possible to easily achieve both objectives. Such situations don't only result to changes in objectives as the project gets on the way but also results to concerns and changes in technical environment or personal and professional aspirations.

Recruitment, Selection and Dismissal

All staff being recruited to work with VLITPs must go through a process that is quite often overseen by the personnel department of the host organization, although the final decision to select a particular candidate is made by one or more members of the project team. There are absolute standards in the remit of the personnel function and the project team may have its own operational standards for all new staff coming on board. Why then is the personnel department involved in the process? The personnel department must not just see to it that there is a person specification against which the required qualities and attributes will be drawn but also ensure there are equal treatment and opportunity in the process, which usually means there is a lack of discrimination due to gender, disability, race or acquaintances (Maslow, 1943). They should insist that all applicants are treated fairly in the interview process, which may involve some form of testing or examination. This is usually part of a recruitment policy within the host organization, which is given to the management team of a VLITP that must be adhered to. The recruitment policy document often contains: how to find the balance in recruitment activities, when to do internal and external recruitment, the need to give training and development to all staff in the organization and what the organization considers a fair remuneration at different organizational levels. Recruitment process should be consider important for the successful implementation of VLITP because getting it wrong often means the implementation could be extremely disruptive and the project could incur unnecessary expenditures or even failure due to lack of competent staff that are definitely in scares supply at the beginning of the 21st century.

The proper understanding of what each job requires when the project team is about to employ someone into a position is included in a requirement for vacancy and takes into account the personal qualities, professional qualities and expertise needed in the individual being sought (Maslow, 1943). Because the key to having productive and effective staff on the VLITPs depends on the selection process, certain standards are used which may include: physical attributes, attainments, general intelligence, personality and temperament, aptitudes, interests, disposition and the state of affairs in the individual's life/career at this stage. While most of this information about the applicant could be obtained from reviewing application form or resume, few would need the applicant to attend an interview or one of the varieties of personality, skill, dexterity, and aptitude tests in use today. All these processes are meant to bring out qualities suitable and relevant for the work required by VLITPs. In cases where the basic skills required for the specific job are not in great supplied, the management team often considers the presence of positive attitudes and commitment in the candidate as important as experience and expertise and increasingly paying very little attention to candidates who give the impression of being a potential celebrity.

On rare occasions the management team of a VLITPs may decide to terminate the services of an employee—most likely for committing serious offences. The major concern of HR here is to ensure that a particular dismissal was fair, unfair or wrongfully carried out. All HR managers would verify where it has become necessary to dismiss an employee, a reasonable, sustainable and supportable procedures and statutory obligation have been followed (Maslow, 1943). The involvement of personnel department in dismissal is to avoid situation where anyone working on VLITPs is dismissed either for no good reason or where a fundamental breach of the employment contract on the part of the employer has occurred. It must also be understood that management teams for VLITPs maintain the sanction of dismissal as the final solution to disciplinary problems. Before dismissal is contemplated, the individual affected must be given the opportunity to state his or her case. Dismissal can also occur when constructively where an individual could be invited by the management team to resign instead of being discharged. This approach should only be applied during a VLITP when an individual working on a given task proves objectionable with significant disadvantage to the work pattern, and result in the individual's continued employment untenable.

When an individual has been dismissed, or when the contract comes to the end, it is against the law for an employer to place any restriction on where they may go to work next. This means when the wrong person is dismissed during the critical period during the implementation of a VLITP, the host organization may just be giving away the secrets of its most ambitious innovation to competitors—another reason why many HR departments of host organizations are unhappy with the increasing decisions by VLITP managers to pay an independent IT recruitment agency to deal with this whole process of recruitment, selection and dismissal.

Induction and Training

Every staff working with VLITPs should go through a formal induction process to realize the importance of being acquainted with the host organization, its policies, practices and organizational objectives. Such understanding should cover how the individual's job fits into the host organization's overall objectives and the importance of implementing that particular VLITP. Induction also provides new staff with an opportunity to get acquainted with the names and status of people on a VLITP as well as the host organization but also working procedures, remuneration and promotion policies, welfare and recreational facilities (Hitt et al, 2000; Morris, 1998). The management team of VLITPs must emphasize the need for induction after an individual has been given a new post within the same sub-project or another sub-project. This should also be done when the host organization has changed management either through takeover or merger (Eisenhardt, 1996). Working with VLITP requires exact knowledge at an early stage where one stands in this new—and often quite temporary—function within the host organization.

The VLITP management also needs to be acquainted with the training policies of the host organization and whether the individuals on the project are entitled to personal training, professional training or organizational specific training at the expense of the host organizational (Forsberg, Mooz and Cotterman, 2000). The types of training usually offered to individuals working with VLITPs are: craft training, software operational training, day release for training, professional training, skill and aptitude training and continuous professional development. It is absolutely essential to train all staff adequately because some might not have been introduced to new technologies involved in this project. However, training activities at various levels must be based on:

- Need assessment and identification.
- Planned and structured programs.
- The monitoring of particular training activity effectiveness.
- Review and evaluation of the success or failure of each training activity.

In certain cases staff may decide to take up a position within a VLITP based on the condition of training program during the implementation period.

Remuneration and Job Evaluation

The aim of remuneration is primarily to reward an individual for productive effort, expertise and output using a specific means of measurement. Not only must the individual feel that the remuneration provides an adequate level of income on a regular basis but also that the individual must be motivated, encouraged and it is a reasonably good incentive for the work he or she has done. Remuneration can be based on achieving targets, often set to meet specific objectives within a specific time limit. Certain categories of work within VLITPs (i.e. secretaries, administrators, receptionists, etc) have problems with these (performance-related pay) schemes because the criteria against which performances are measured is often not cleared or fully understood and, more importantly, does not depend solely on their own efforts (Hitt et al, 2000). Nevertheless, VLITPs adopt a combination of reward strategies for different categories of staff to cover a variety of aims and purposes in response to particular situations. While it might seem very difficult for any VLITP to be totally fair to all staff at all times, those responsible for making policies and implementing remunerations must take steps to ensure that fairness can be achieved more often than not.

The goals of VLITP managers in evaluating nearly every job within the VLITP are to properly assess the nature of work involved and the task being undertaken by workers in different sub-projects. The evaluation exercise also helps to find a balance between difficulty, value, frequency, importance and contribution of a particular task on the project. It assists the subordinate in understanding/clarifying the tasks to be carried and the marginal or infrequency of the tasks being measured against a number of criteria. These criteria could be: whether the skills, qualities and attributes matches the specified level within or outside the host organization. It also helps to match up jobs against factors like grades, job titles, or salary scales—within a process of revolving anomalies and differences that is considered speedy enough by all parties.

Health and Safety

Together with the cooperate responsibility for ensuring that an overall healthy and safe working environment is provided to all individuals working on VLITPs, the managers of sub-projects are directly responsible for all matters related to health and safety within their project teams (Payton, 2003). A health and safety component must be an integrated in the induction and training stages for all staff on VLITPs. The management team must also ensure that safe behavior patterns are devised to support the required attitudes and give effect to the health and safety procedures. It is very critical that all staff understand the need for any potential hazard in any part of a VLITP domain to be dealt with by whoever happens in awareness of danger. Not only is it a legal requirement in most countries today that all accidents are reported immediately, but this also provides the project manager with relevant information about accidents in the work area that need to be analyzed for relevant steps to be taken to avoid future recurrence (Payton, 2003).

Staff Discipline

When the concentration of all is deeply focused on the achievement of a particular project goal within a specific period of time, there is a need to have rules of behavior that are accepted and being obeyed by all for staff. Most common among such rules are punctuality and absence, as well as the qualities of output and performance of all project staff. Quite often codes of discipline are written into staff service contracts, usually adopted from the host organization staff handbook (Payton, 2003). All new staff joining any part of VLITPs must be made aware of the code of discipline and the need for them to comply as a key part of an induction process. This sets the scene for this code to be referred to in the rare occasions that it may be necessary to discipline staff on a VLITP. It is a good practice that an individual who is faced with disciplinary actions during the implementation of VLITPs be entitled to be accompanied by a work colleague or representative from a trade union. They should also be entitled to face any accuser and be given an opportunity to state their case. Even when a decision has been reached the individual still has an opportunity to appeal against the outcome of the investigation.

VLITP BOUND AND UNBOUND

Despite their differences, instrumental and substantive theories share a similar attitude—take it or leave it—toward IT projects. On the one hand, if IT projects are mere instrumentality, indifferent to values, then the designs and structures of VLITPs are not issue in public debate, only the effectiveness and efficiency of the emerging applications (Currie and Guah, 2007; Gershon, 2004). On the other hand, if VLITPs are the vehicles for a particular organization's domination in an industry, then the public will condemn either to pursue the advances of such projects toward dystopia or to regress to a more primitive way of during business. In either case change can be disputed: in both theories, *IT projects are* destiny. Reason, in its technological form, is beyond the public intervention after the initial decision to implement has been taken by the host organization.

This accurately justifies why most evaluation of failed VLITPs seek only to place a boundary around them, rather than transform them. The case of NHS national program (see Chapter XIII) have been widely criticized for spending much needed money for bedside nurses and penetrating "too far" into zones where medical doctors should have complete control. Automatic systems within the airline industry are under constantly attack by staff of on grounds of quality of service to customer and airline safety (see Chapter XIV). In both cases critics urge the public to reject certain VLITPs, and then ask the public to accept the price of preserving traditional ways that keeps face-to-face interaction between customer and staff. This agenda has given rise to both moral and trade union solutions to the problem of implementing VLITPs.

Moral boundaries: While trade union bosses in the host organizations may seek to reinvigorate institutions (such as the family) on a traditional basis, individual staff focuses on consequential values. Certain existing work have condemned the implementation of VLITPs for reducing employees to mere instrument for the achievement of wealth and power, and call for a restoration of the total value of employer-employee commitment (Haag et al, 2006). Progressives worry about the subversion of tradition institutions by VLITPs. Some have argued that working with multinational organizations presupposes a commitment by the staff to engage in rational argument. The contention is that multinational organizations technologize the working environment by transferring staff functions to robotic experts and destroy the very meaning of employment.

Cleland (1990) offers a sophisticated version of the idea of a return to simplicity—with no implementation of VLITPs. This, however, calls for a "two-sector" economy in which an expanding craft sector will take up the slack in employment from an increasingly automated economic core. This view is premised on an uncritical acceptance of the dominant technological paradigm which, Cleland (1990) asserts, "is perfect in its way". But are all VLITPs really "perfect" in conception and design? Do they not rather lead businesses further into disaster? And how can the host organization confine this disaster to its unique sphere in relation to the wider industry, as all these theorists suggest, when the problems it creates overflow every boundary and shape the whole framework of business arena?

In an effort to put some order in this barrage of objections, these four reasons have been compiled for doubting that moral solutions will stance any chance against VLITPs:

1. While the view of technical progress in a modernized organization does not necessarily mean imperialism, implementing very large a IT project can be regarded by staff of the host organization as relative to other dimensions of further controls. It is just as important to conceptualize the progressive transformation introduced in the host organization as a result of VLITPs by defining its limits. All too often, having defined the objectives of VLITPs in their proper place, criticism fails to see the potential and, in condemning the project's current form, can also serve to foreclose the VLITP's possible future.
2. Suppose, however, that one succeeds in combining limits on a VLITP's reach with an effort to reform the emerging technology within its own domain. The problem still remains of defining that domain. It is extraordinarily difficult to reach agreement on which existing activities in the host organization should be protected from invasion of the VLITP. The only consensus values left in modern management strategy are effectiveness and efficiency—precisely the value critics of VLITPs are attempting to bound so that other values may flourish.
3. On the other hand, by placing trade union members' values in rigid opposition to implementation of VLITPs, the host organization concedes what needs

to be defended, i.e. the possibility of a technically rational civilization that enhances rather than undermines those values they are committed to uphold. The moral critique of VLITPs always seems to reopen the tedious debate over "principles" vs. "profit motives." In a democratic society there is no debate but a confession of impotence, since the victory of the profit motive is so very predictable. What is needed is an alternative practicality more in accord with principle. That is what traditional Marxism promised, but failed to deliver. The question being posed as the 21st century unfolds is whether today business executives run their organizations any better without implementing VLITPs.

4. Finally, the very effort of bounding VLITPs appears suspect. If team leaders *choose* to leave something untouched by VLITPs, is that not a subtler kind of technical determination? If the host organization feels the sudden need for meaning in this overly technologized business environment, and obtain it by returning to previous business traditions, are they not *using* tradition as a kind of super IT project? If so, how can the trade union members accept it? How can multinational businesses ever leave the technical sphere if the very act of bounding a VLITP instrumentalize them?

Trade union solution: The trade union solution to the problem of bounding VLITPs turns out to be no more promising. Certain Asian and Eastern European countries that attempt to preserve indigenous values while modernizing technically have tested this solution. Typically, the trade union organizations have argued that the flaws of modern society are the result of a specific instrumentalization of VLITP. The view that American form of modernization and its peculiar techno-culture activities as a system of "values" represent that something foreign to real European way of running business? Their goal is of building sub-regional economic and cultural spheres, fits in quite well with the gradual improvement process of business management, where small to medium sized IT projects are in the service of these alternatives.

Though the differences are cleared, the Russian experience resembles that of Japan except that the Russian Revolution was oriented toward values that would be realized in the future rather than toward values from the past. Once again, the protection of these values required the energetic acquisition of existing technology to achieve rapid economic development. Despite certain substantivist implications of the Marxist theory of economic stages, the then regimes in Russia adopted a typical instrumentalist position on implementing technological projects—often using and importing them as though they were neutral tools. Tight control of economic and cultural interaction with the capitalist world was supposed to open a protected space within which a new culture would be born. The loss of cultural control is so complete that no turning back seems possible. Reviewing the Russian experience one understands the current difficulties to believe in the rearguard defense of cultural isolationism in China in the context of intensified economic exchanges with the West.

Instrumental theory of implementing VLITPs is not entirely refuted by these experiences, although in each case governments were unable to use technology to further original cultural goals (Tolbert and Zucker, 1983). Defenders of the instrumental view sometimes draw comfort from the conjunction of democratic reform with the decision for modernization. The general public, on the other hand, appear to have refused the trade-offs required to sustain traditional or future-oriented values in competition with well being in the present. The conquest of business by VLITPs is not due to any occult power of the "technical phenomenon"; rather, implementing VLITPs, as a domain of perfected instruments for achieving improvement in products and services, is simply a more powerful and persuasive alternative than any ideological commitment.

At this point the specificity of the instrumental theory collapses. If VLITPs are truly neutral, they should be able to serve a plurality of ends. But the close association of mass democracy with modernization seems to deny that pluralism, and in fact confirms the arguments of substantive theory (Tolbert and Zucker, 1983). There is little reason to distinguish the two theories if they disagree only in their attitude toward various emerging project objectives foreseen by both.

A more interesting argument divides the substantive approach from Marxist critical theory. Both can agree that the Japanese and Russian examples differed only superficially from the American way of doing business they professed to transcend. Substantive theorists see this as evidence that no alternative technological modernization is possible. But critical Marxism argues, on the contrary, that an alternative may yet be created on the basis of trade union control, re-qualification of the labor force, and public participation in top management decisions to implement VLITP. Describing the Japanese and Russia experiments as failure is partly because they rejected the latest business management strategy path for one convergent with authoritarian industrialism.

According to this view, the attempt of states to instrumentalize VLITPs on behalf of original values represents an internal contradiction. In the face of the challenges to implementing VLITPs (mentioned throughout this book), only a particularly strong state can create a culturally and economically closed region for the furtherance of original cultural goals. Yet paradoxically a strong state can only sustain itself by employing the authoritarian technical heritage of business management. In so doing, it reproduces all the main features of the civilization it professes to reject: predictably, the means subvert the ends.

DEALING WITH TIME LIMITS

Time management is a critically important skill for any successful project manager. Several researches have reported that project managers who succeed in meeting their project schedule have a good chance of staying within their project budget. The most common cause of blown project budgets is lack of schedule management. Fortunately there is a lot of software on the market today that contributes to proper management of project schedule or timeline.

Simple IT projects (see Chapter I) typically have few tasks dependent on other tasks, and will be relatively simple and easy to coordinate. They can therefore use simple project management tools (like Gantt Charts) to handle time limits. Simple projects are often best run using simple timetables and action plans. These should be prepared and negotiated with project staff to improve plans and get buy-in. Critical Path Diagrams may overcomplicate project scheduling and communication for such projects. A common problem with the IT industry is failure to properly train project team members adequately in the use of these simple tools. Instead, project managers can often 'blind people with science', leading to poor communication and muddled projects.

Appropriate timetables and action plans are often sufficient to coordinate and implement simple projects. These should be explained and negotiated with project staff to improve the plans and get staff understanding, input and buy-in. It will often be enough to create a work-back schedule, starting from the date by which the project must be completed, and listing all of the tasks in reverse order with due dates for each.

Team leaders of sub-projects must ensure there are agreed scopes with the host organization before the planning phrase begins. This will help the project managers to resist changes to its scope (known as "scope creep"), which will seriously affect their plans, once they have started working. During the project these will contain sufficient control points and deliveries to monitor project progress and take any appropriate remedial action.

But how do team leaders ensure that they really have covered everything? Would anyone else evaluating certain aspect of the overall project implementation strategy, at a later date, know the progress with the project if the managers were unexpectedly off sick for a few days? And are the project managers quite clear about when they need to start if everything is to be done and dusted by the deadline? For these and other reasons an Action Plan for all sub-projects are needed, which is a simple list of all of the tasks that team leaders need to carry out to achieve the project objective. It differs from a To Do List in that it focuses on the achievement of a single goal of a sub-project as part of a VLITP.

Wherever project managers want to achieve something significant, they draw up an Action Plan. This helps them think about what they need to do to achieve that thing, so that they can get help where they need it and monitor their progress. To draw up an Action Plan, they simply list the tasks that are needed to achieve the goals of the VLITP. This may seem very simple, but is still very useful. Action Plan must be kept close by as the system is being implemented and update it as they go along with any additional activities that come up.

When there might be a need to try and achieve a similar goal in the future, it is advisable to revise the Action Plan after the work is completed, by changing anything that could have gone better. It gives information about projects that could have avoided on a last-minute panic if suppliers can be alerted in advance about when and approximately what size of order they would be placing.

Action Plans are great for sub-projects, where deadlines are not particularly important or strenuous, and where team leaders don't need to co-ordinate people from other sub-projects. An Action Plan is a list of things that project managers need to do to achieve a goal. To use it, simply carry out each task in the list.

EFFECTIVE AND EFFICIENT UTILISATION OF RESOURCES

A VLITP is likely to miss the schedule date as a result of underestimating the length of time or the amount of resources required for various tasks, but omitting a task in a VLITP could lead to an incomplete or unaccomplished outcome. Avoiding mistakes in the sequencing of project tasks can ensure VLITP meet schedule. Quite often VLITP managers build the project schedule by orderly listing all the tasks needed to be completed. Each task is assigned an estimated duration and appropriate allocations are made for the required resources (Gershon, 2004). Determine predecessors (what tasks must be completed before the next) and successors (tasks that could not start until the previous one is completed) each task. It helps to think of this in a pretty simple and straightforward manner. For instance, like a project called "Getting Ready To Go Out". The task "put on shirt" may have a longer duration if it is a buttoned dress shirt than if it's a pullover. It doesn't matter which order you complete the tasks "put on right shoe" and "put on left shoe", but it is important to complete the "put on pants" task before starting the "put on shoes" task.

The difficulty in managing a VLITP schedule is that there are seldom enough resources and enough time to complete each task sequentially. Therefore, several tasks often have to overlapped thus taking place simultaneously. VLITP managers can greatly simplifies the task of creating and managing the project schedule by handling the iterations in a scheduled logic. Listing all tasks, allocating resources, and then sequencing them to see that some tasks have a little flexibility in their required start and finish date is referred to in VLITP as 'floating'. Other tasks have no flexibility—referred to as having a 'zero float'. A line through all the tasks with zero floats is called the 'critical path'. All tasks on this path (and there can be multiple or parallel paths) must be completed on time if the VLITP is to meet it schedule. The key time management task for VLITP managers is to manage such critical path.

It must be noted here that items can be added to or removed from the critical path as circumstances change during the execution of a VLITP (Gershon, 2004; Morris, 1998). Relative major issues—like the installation of security cameras—may not be on the critical path, but if the shipment of certain raw materials is delayed, the security of existing materials may then become part of the critical path. Conversely, customizing an of-the-shelf package may be on the critical path, but if the software supplier releases a new version of the package for the environment that exercise could come off the critical path (or reduce the length of the critical path). Regardless of how well one manages a VLITP schedule and the resources, there is one more critical element—managing the budget (Patel and Morris, 1999).

Some of these estimates will be more accurate than others. The host organization often has a reasonably good idea what it will charge to each of its sub-projects for different classifications of labor. However certain commodities can be priced in a very competitive market resulting to prices that can be fairly unpredictable. Thus, certain estimates are less accurate. For instance, the cost of a system testing with higher performance specifications that normally can be estimated to be more expensive, but it is hard to determine whether it will be 10% more or 15% more. For an expensive item, that can be a significant amount. When the estimated cost of an item is uncertain, the project budget often includes a design allowance (Hitt et al, 2000; Weill, 1993). This is means money has to be set-aside in the budget "just in case" the actual cost of the item is wildly different than the estimate. Unexpected political events—September 11, 2001 attack on the World Trade Center in New York—or unsuspected problems with suppliers are always a possibility on VLITPs. The host organization usually includes a contingency amount in the project budget to cover these kinds of unforeseen events.

As a precaution all VLITPs budget is composed of the estimated cost, plus the contingency and design allowance, plus any anticipated profit. A major function of the team leaders of sub-projects is to keep the actual cost within the range of the estimated cost. The usual strategy is to try to use as little of the design allowance and contingency budget as possible and endeavor to maximize the profit the host organization earns on the project. To maximize chances of a VLITP meeting its budget, the project schedule must be met—the most common cause of blown budgets is blown schedules. While meeting the project schedule doesn't necessarily guarantee meeting the project budget, it significantly increases the chances. Even more critical for success is managing the VLITP scope by not permitting the project scope to "creep" upward without getting budget and/or schedule adjustments to match.

OUTSOURCING VULNERABILITY

The habit of management teams to purchase (outsource) certain parts of VLITPs to an IT service provider has not only grown in status in the 21st century but has also beginning to be relied upon as a major contribution to the success, effectiveness and economic solution to nearly all VLITPs (Lacity and Willcocks, 2006). This is where the managers of a VLITP decides to hire the services of another organization—usually an IT service provider or Outsourcing company—to supply particular software, component, application or service feature for the successful completion of the project (Guah, 2008). This type of purchasing can have either positive or negative effect on the finances of VLITPs. However, it is widely considered to be a variable cost to every VLITP and the service providers promote themselves as making an enormous contribution to the successfulness of the project through their capabilities in keeping such costs down, while ensuring the reliability and quality of the individual component being supplied.

To successfully accomplish the project objectives, using outsourcing companies or IT service providers, the managers of VLITPs need to negotiate and manage the financial situation to ensure regularity of the system components in question to have continue access to other sources (Currie and Guah, 2006; Hitt, 2000). They can do this by applying several purchasing activities during the implementation of VLITPs. The purchasing team negotiation in the interest of the VLITP must have a wide knowledge of the markets within that particular area of the IT industry. Such knowledge should extend to how the suppliers perform in regard to reliability, staffing situation and previous delivery performance (Patel and Morris, 1999). The purchasing team will also need to liaise closely with the IT service provider ensure the is no risk of the supplier not keeping to delivery undertakings and the necessary requirements would be readily available (Gershon, 2004).

A common practice during the implementation of VLITPs is that the purchasing team will be responsible for negotiating purchasing prices and terms as well as ensuring that the components of the project ordered and invoiced are actually receive and the appropriate prices are paid. Procedures are often put in place by the purchasing team to avoid fraud at this stage of managing a VLITP. The purchasing team continuously monitors expected deliveries to ensure that service providers adhere to their undertakings regarding quality and delivery dates.

There are several approaches a purchasing team can use to deal with IT service providers and outsourcing companies. These are few of the common approaches: by now pay later, pay now buy later, discount preferred supplier, just-in-time, retainer systems, future and international financial instruments, extended credit and debit schemes, etc. The aim here is to acquire the right component of the project, using the right expertise, with the right information, at the right time, delivered to the right place, and on the right terms.

Like other aspects of VLITPs, purchasing of project components from IT service providers must be done according to a specific policy. Considering such policy is in support of the wider objectives of the goals of the VLITP and the host organization as a whole, the chief executive and top management should focus on the policy seriously:

- One important element of the policy to guide the purchasing team is the level, critical nature and size of the total project to be contracted out to IT service providers to completion outside the main IT project team.
- Another important element of the policy to guide the purchasing team is how much involvement to have with speculative buying.
- A significant element of the policy to guide the purchasing team is whether or not the purchasing team can enter into short-term or long-term contracts and whether the purchasing team should be committed to one or more IT service providers during the implementation of the project.
- Probably the most debatable element of most policy to guide the purchasing team is giving the team a choice to retain expertise and information access, especially at the beginning of the 21st century when there is scarcity of specific expertise are in very high demand and short supply, even to IT service providers.

The consistency and reliability of these purchasing policies within VLITP are much more improved when the purchasing team utilizes large scale and relatively uniform centralized purchasing strategy. That's because this strategy often results in consistent buying policy with fewer problems for both purchasing team—with more power to influence service providers in regard to prices and terms with a large single order—and the service providers—inclined to give better and more favored attention to any organization placing large for fear of losing such a valuable customer. These advantages of centralized purchasing only serve to compensate for the considerable loss of flexibility and occasionally being forced to settle for a slightly less than ideal component of the VLITP. On his note, great efforts should be made to avoid a situation that may necessitate the alteration to the specification of the overall project by the managing team which may prove in the long term to be more damaging or add more expense to the total costs of a VLITP (Patel and Morris, 1999). Due to the enormous nature of VLITPs central purchasing can become unwieldy and bureaucratic leading to delays in authorizing and placing—often to the detriment of the work of project team members at the lower level whose work may be complimentary to the outsourced component of VLITPs.

As an alternative to the strategy of central purchasing, a purchasing team may decide to give each project manager of a sub-project within a VLITP the control over their component inputs thus bringing a greater level of operational experience to bear on the final selection and acquisition. While this strategy may lead to an increase in component costs, an inability to bargain from the strength of large orders and the arguably a greater opportunity for fraud, the actual orders involve significantly less bureaucracy, are placed more quickly and followed up more effectively (Benveniste, 1987).

PROJECT TEAMS

The success of VLITP managers in attaining project's objectives can only be done with the willing co-operation of all member of the project team. The need to lead from the top is absolute for staff at all level to be motivated to co-operate in the interests of meeting the project's objectives (Benveniste, 1987). Such need to lead from the top does not necessarily mean the promotion of bureaucracy at different levels of the project. Instead, there should be a decisive and relevant leadership teams managing the various sub-projects with the capacity to motivate the rest of the team to work efficiently. The size and complexity of VLITPs makes it impossible to always manage them democratically. Each member of staff often has different requirements, ideas and agendas. Trying to please everyone on the sub-teams can lead to project paralysis.

An even important member of the VLITP team that probably needs the most attention and involvement is the end-user. They need to be kept on board all through the project and not just be given a blank piece of paper at the beginning asking them to define what they want (Benveniste, 1987). Keeping the end-user requires VLITPs that consist of a small team of 'doers' who can make things happen. Such team needs to be operationally biased because they are the people at the sharp end who are going to do most of the work during the implementation and who are going to need to reorganise themselves to achieve the benefits. They also need full and genuine input into the management of the project. Quite often the in-house IT department finds it difficult to cope during the implementation of VLITPs simply because they are there to support the host organization and not to manage new projects (Smith and Willcocks, 1995). Stuffing this department with support staff inevitably creates discord and marginalizes the entire project.

Table 7.1. Example of project management team

Successful team	Failed team
Deputy chief executive/FD	Finance director
General manager surgery	Information manager
General manager medicine	IT manager
Information manager	Hospital's project manager
Medical records manager	Medical records manager
Clinician	Contracts manager
Supplier's project managers	Training manager
+ Seconded Staff when required	Medical director

Table 7.1 shows two examples of teams we discovered during our research who were part of the implementation of a large IT projects in a European hospital. The successful team in the left column was described as the core implementation team at this hospital. The failed team in the right column was known to the hospital as a procurement team responsible for purchasing a hospital-wide clinical information and electronic patient record system. It was also noticed from our research that the service provider supplying the system was very much part of the successful team and that the majority of team members were taken from operational rather than support services. On the other hand the failed team consisted of top executives from different parts of the hospital who decided to ploughed on for a few years without achieving anything until there was a clinicians' revolt in the third year which led to a complete restructure. The failed team was later made to consist of four consultants, the information manager and the operations director. This new project management team was still not ideal as there was no finance input. It took a few incarnations and some tears to get the composition of this group right, but the team has been working extremely well for a year up to the end of our research.

Table 7.1 shows two examples of a project management team setup to implement an IT project in the medical institution in Europe:

These two examples show that the VLITPs team should:

- Not only lead from the top but also avoid bureaucracy.
- Be capable of being decisive.
- Consist of relevant management members.
- Also be empowered to manage the necessary procurement using a well-respected procurements methodology (Smith and Willcocks, 1995).

The new 21st century work a habit has made it increasingly difficult for financial rewards alone to motivate staff working on VLITPs to given their maximum effort. As a result VLITPs manager are increasingly confusing the two: motivation and incentive. It is important to understand the difference between these two. They are intrinsically different mainly because motivation springs from within while incentive belongs to those forces that are applied from outside (Payton, 2003). Motivation is essentially an attitude of mind that may be encouraged by external factors but is fundamentally firmly related to self-discipline. Incentive is a positive external influence to encourage improved performance (Payton, 2003). Motivation can remain even when the external influences that engendered it have ceased to exist because it is self-generated arising through innate character attributes and so can remind unaltered by external factors (Proenca, Rosko and Zinn, 2000). But the effects of incentives cease as soon as those particular influences are withdrawn.

The above difference justifies the need to have staff working on VLITPs that are motivated by means other than just incentives that ate quite often only financial. VLITPs managers should therefore realize that people working on sub-projects do not work for money only. While the reasons to work are different for different people, working on any particular VLITP—and so are the factors that motivate them—several research work have shown that once a satisfactory income has been achieved, financial incentives become a secondary consideration even though it may be given emphasis at different times in the career (Eyre and Pettinger, 1999: 127-132). There are many other ways to motivate staff on VLITPs. A job that gives the staff a status in their own environment both socially and at work may proof to be a good way of motivating some staff. Others could be motivated by the security of a job spell that would last for a specific period. While some may prefer to continue in one job or organization for a very long period (because they like working within familiar surroundings) others may like to move around often for the challenge of a new and different environment. Some people working on VLITPs may also find dull and repetitive work to be less motivating because they are proud of their skills and really want to employ them to the best way possible. Nearly all staff on VLITPs are much more motivated when they know they are working within a positive environment than a negative environment with bad adversarial managerial styles. It can be argued that everyone working on VLITPs has individual goals and aspirations that for some can be fulfilled through their work while for other those goals and aspirations can only be accomplished elsewhere outside their working lives. Generally, all staff of VLITPs believes in the ambitious nature of these projects. They are therefore motivated by the fact that VLITPs bring them the chance to respond to the opportunities to progress, develop and enhance their range of skills, qualities and experience (Guah, 2008). Many people working on VLITPs are motivated by the fact they are working in the most well respected industry with fast development and changes which often reflect on their own abilities and the status of their knowledgebase. Finally, management teams of VLITP must accept that all members of the project team require to be treated with respect and be shown that they are valued. All staff must understand that they are being paid fairly for the work that they are doing in comparison with other both within the outside that particular organization.

The motivation of individuals can either be positively or negatively affected by the team they belong to (Eisenhardt, 1996). Because VLITPs are implemented by stages and division of objectives handled by smaller teams of project staff, the issue of team motivation is just as important as individual motivation. There are certain areas that should be looked at for a team to be motivated successfully. Not only should every member of a team within a VLITP be subject to the same standards of discipline, attendance, output and control, the value of their contribution must also rewarded equally within the team. This does not necessarily mean that everyone in a team should be on the same salary but the pay should be based on fair level of skills and responsibilities within the sub-project. Likewise opportunities should be offered on equal basis with every member of the team feeling they have been treated fairly at all times. This involves the distribution of work amongst members of the team based on fair principles with praises for the team achievements and blames for the team mistakes shared equally to all members. This avoids the situation where a single individual on the team is consistently singled out for special treatment whether positive or negative (Feldman, 2004; Proenca, Rosko and Zinn, 2000). By ensuring that team supervisor/manager sets reasonable and achievable targets for all members, the achievements of the team as a whole can be recognized and appreciated as well as proclaimed to the team as whole—often during occasional team social activities.

CHAPTER SUMMARY

All VLITPs have specialist personnel or human resource management functions to deal with the matters that have discussed in this chapter. Sub-teams quite often also have one individual capable of handling these issues and may use the services of experts when required. Due to the vast and varied scope of human resource management issues today, all VLITPs will be involved in most these activities at some stage of the project implementation.

People working on VLITPs have various reasons with a different combination of attitudes and motivations. However a recognition that the reasons listed in this chapter exist should be very helpful to the management teams of VLITPs in driving motivation of their staff at various levels for ultimate aim of achieving the project objectives successfully.

The purchasing team constantly investigates the particular area of IT industry seeking better services, prices as well as terms of deliveries (Feldman, 2004). In the process the purchasing team seeks out alternative service providers to ensure the continuous delivery should the principal service provider fails to deliver for any reason. This strategy also ensure there is a multiplicity of service providers to maintain competitiveness in the IT industry. By being aware of the prevailing state of the service provision market in different areas within the IT industry, the purchasing team can only anticipate problems but also take proactive measures for the successful completion that project.

A good organization should enlarge the personal freedom of its members while enabling them to participate effectively in a widening range of business activities. At the highest level, being a business executive involves choices about what it means to run a business the best way possible. Technological developments in the 21st century make these choices increasingly mediated by the implementation of VLITPs. What level of business activities organizations are and will get involve is decided in the shape of IT projects no less than in the action of share holders and board members. The chose of objectives for VLITPs are thus ontological decision fraught with serious business consequences. The exclusion of the vast majority from participation in this decision is the underlying cause of many of VLITPs failure today.

REFERENCES

Benveniste, G. (1987). *Professionalizing the Organization: Reducing Bureaucracy to Enhance Effectiveness*, Jossey-Bass.

Cleland, D.I. (1990). *Project Management: Strategic Design and Implementation*, TAB Books, PA, p-23.

Currie, W. L., & Guah, M. W. (2006). Web Services in National Healthcare: The Impact of Public and Private Collaboration. *Information Systems Journal*, *1*(2), 48-61.

Eisenhardt, K. M. (1996). Resource-based view of strategic alliance formation: Strategic and social effects in entrepreneurial firms, *Organizational Science*, 7(2), 136-150.

Feldman, M. S. (2004). Resources in Emerging Structures and Processes of Change. *Organization Science, 15*(3), 295-309.

Forsberg, K., Mooz, H., & Cotterman, H. (2000). Visualizing Project Management, Second Edition, Wiley, pp. 89.

Gershon, P. (2004). *Releasing Resources to the Frontline: Independent Review of Public Sector Efficiency.* Her Majesty Treasury: London, July.

Guah, M. W. (2008). Changing Healthcare Institutions with Large Information Technology Projects. *Journal of Information Technology Research*, **1(1), 14-26.**

Haag et al. (2006).

Hitt, M. A., Dancin, M. T., Levitas, E., Arregle, J. L., & Borza, A. (2000). Partner selection in emerging and developed market contexts: Resource-based an organizational learning perspectives. *Academy of Management Review*, 14(4), 532-550.

Lacity, M. C., & Willcocks, L. P. (2006). Transforming back offices through outsourcing: Approaches and lessons. In Leslie P. Willcocks and Mary C. Lacity *Global Sourcing of Business and IT Services*, Palgrave Macmillan, pp.97-113.

Markus M. L., & Robey, D. (1988). Informational technology and organizational change: Causal structure in theory and research. *Management Science*, 34(5), 583-594.

Maslow, A. H. (1943). A theory of human motivation. *Psychological Review, 50*, 370-396.

Morris, P. W. G. (1998). Key Issues in Project Management. In Project Management Handbook edited by J. K. Pinto, Jossey-Bass, p. 5.

Patel, M. B., & Morris, P. G. W. (1999). Guide to the Project Management Body of Knowledge, Center for Research in the Management of Projects, University of Manchester, UK, p. 52.

Payton, F. C. (2003). E-Health Models Leading to Business-to-Employee Commerce in the Human Resources Function. *Journal of Organizational Computing and Electronic Commerce, 13*(2), 147-161.

Proenca, E. J., Rosko, M. D., & Zinn, J. S. (2000). Community Orientation in Hospitals: An Institutional and Resource Dependence Perspective. *Health Services Research.*

Smith, J., & Willcocks, L. (1995). IT-enabled business process reengineering: Organisational and human resource dimensions. *Journal of Strategic Information Systems*, 4(3), 279-301.

Tolbert, P. S., & Zucker, L. G. (1983). Institutional sources of change in the formal structure of organizations: The diffusion of civil service reform, 1880-1935. *Administrative Science Quarterly, 28*, 22-39.

Weill, P. (1993). The Role and Value of IT Infrastructure: Some Empirical Observations. In *Strategic Information Technology Management: Perspectives on Organizational Growth and Competitive Advantage.* M. Khosrow-Pour and Mahmood, M. (eds.). Hershey: Idea Group Publishing, pp. 547-72.

Chapter VIII
Ergonomics of Very Large IT Projects

ABSTRACT

Medical accidents, such as those that occur as a consequence of errors in medical systems, rarely happen because of a single failure. They are usually the consequence of a multiple breakdown in the system. This chapter explores the potential for risk and, demonstrates the need to improve design interventions in a VLITP context. It considers issues that range from the design, packaging and labelling of VLITP environment in which medical systems error might occur. The ergonomics systems approach to VLITP is an appropriate method for involving all key users and for addressing their needs. This requires generic issues to be considered. Such issues include: task design, interface analysis, communication interface, variation in user characteristics, and needs (including motivation and culture), training needs, work organizational issues, and the evaluation of interventions and current practice.

INTRODUCTION

Ergonomics is a relatively new branch of science which has been practiced for about 6 decades, but relies on research carried out in many other older, established scientific areas, such as engineering, physiology and psychology (Buckle et al, 2003; Cambridge, 2003; Stanton and Stammers, 2008). The role of ergonomics in developing design for VLITP is to determine requirements, recommendations and suggestions for reducing error. Ergonomics originated in World War II, when scientists designed advanced new and potentially improved systems without fully considering the people who would be using them. It gradually became clear that systems and products would have to be designed to take account of many human and environmental factors if they are to be used safely and effectively. This awareness of people's requirements resulted in the discipline of ergonomics.

WHAT IS ERGONOMICS?

Ergonomics is the intense application of 'user compliance' when designing technology. This approach, during VLITP implementation, puts human needs and capabilities at the focus of designing technological systems. The aim is to ensure that humans and technology work in complete harmony, where the equipment and tasks aligned to human characteristics (Stappers et al, 2007). Ergonomics has a wide application to everyday domestic situations, but there are even more significant implications for efficiency, productivity, safety and health in work settings including:

- Designing equipment and information systems that make IT easier to use and less likely to lead to errors in after rolling the system out. This is particularly important in high stress and safety-critical operations such as server rooms.
- Designing tasks and jobs so that they are effective and take account of human needs such as rest breaks and sensible shift patterns, as well as other factors such as intrinsic rewards of work itself.
- Designing equipment and work arrangements to improve working posture and ease the load on the body, thus reducing instances of repetitive strain injuries and other work related disorder in limbs.
- Information systems being designed to make the interpretation and use of instructional materials, leaflets and books, signs, and displays easier and less error-prone.
- Design of training arrangements to cover all significant aspects of the job concerned and to take account of human learning requirements.
- Designing working environments, including lighting and heating, to suit the needs of the users and the tasks performed. This could also involve the design of personal protective equipment for work and hostile environments.
- For people with minimum experience with IT, this may include assistance by enhancing basic technology for easier acceptability and effectiveness.

Figure 8.1. Human-task environment in ergonomic systems

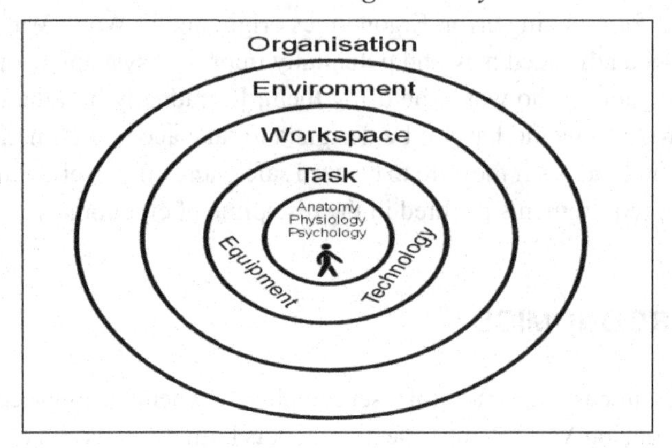

The multi-disciplinary nature of ergonomics (sometimes called 'Human Factors') is immediately obvious (Buckle et al, 2003; Mohamed and Irani, 2004). Ergonomics can be applied in connections with a variety of other professions: design engineers, production engineers, industrial designers, computer specialists, industrial physicians, health and safety practitioners, and specialists in human resources. The overall aim of involving ergonomist in VLITP is to ensure that VLITPs consider existing knowledge of human characteristics and concentrate on practical problems of people at work and in leisure. While many existing systems relay people capacity to adapt to unsuitable conditions, such adaptations quite often lead to inefficiency, errors, unacceptable stress, and physical or mental cost to individuals (Midden and De Vries, 2006). Figure 8.1 shows a human-task-environment, which demonstrates influences human have on performance of a system in the host organization, taking into consideration the systems nature of ergonomics. The final objective of ergonomics is to ensure working in VLITP environments do not result in eyestrain and muscle fatigue or some other human irritations and inconveniences which are currently not inevitable.

ERGONOMICS COMPONENTS

Ergonomics deals with the interaction of technological and work situations with the human being with its origins from several basic human sciences including anatomy, physiology and psychology. These sciences are applied by ergonomics with two main objectives: (1) the most productive use of human capabilities; and (2) the maintenance of human health and well-being. They can both be paraphrased as follows: the job must 'fit the person' in all respects, and the work situation should not compromise human capabilities and limitations.

The contribution of basic anatomy lies in improving physical 'fit' between people and the things they use, ranging from hand tools to aircraft cockpit design (Stappers et al, 2007). Existing literature in human physiology supports two main technical areas. Work physiology addresses the energy requirements of the body and sets standards for acceptable physical work-rate and workload, and for nutrition requirements (Stanton and Stammers, 2008). Environmental physiology analyses the impact of physical working conditions—thermal, noise and vibration, and lighting—and sets the optimum requirements for these.

Psychology is concerned with human information processing and decision-making capabilities (Kaber, 2006). In simple terms, this can be seen as aiding the cognitive 'fit' between people and the things they use. Relevant topics are sensory processes, perception, long- and short-term memory, decision-making and action. There is also a strong thread of organizational psychology. The importance of psychological dimensions of ergonomics should not be underestimated in today's 'high-tech' world—remember the video recorder example at the beginning. The ergonomist advises on the design of interfaces between people and computers (Human Computer Interaction or HCI), information displays for industrial processes, the planning of training materials, and the design of human tasks and jobs (Cambridge, 2003; Kim and Kwahk, 2007). The concept of 'information overload' is familiar in many current VLITP jobs. Paradoxically, increasing automation, while dispensing with human involvement in routine operations, frequently increases the mental demands in terms of monitoring, supervision and maintenance.

Underlying all ergonomics work is careful analysis of human activity. It must understand all of the demands being made on the person, and the likely effects of any changes to these—the techniques which enable him to do this come under the portmanteau label of 'job and task analysis'. The second key ingredient is to understand the users (Kaber et al, 2006). For example, 'consumer ergonomics' covers applications to the wider contexts of the home and leisure. In these non-work situations the need to allow for human variability as at its greatest—the people involved have a very wide range of capabilities and limitations (including the disabled and elderly), and seldom have any selection or training for the tasks which face them. This commitment to 'human-centred design' is an essential 'humanizing' influence on contemporary rapid developments in technology, in contexts ranging from the domestic to all types of industry.

Several situations in the workplace today epitomise the shortcoming in design when no account has been taken of the end users (Kaber et al, 2006; Kim and Kwahk, 2007). Even the standard keyboard on the office PC does not take into account that people come in all shapes and sizes. The ergonomist takes this variability into account when influencing the design process. The branch of ergonomics that deals with human variability in size, shape and strength is called anthropometry. Tables of anthropometric data are used during a VLITP to ensure that places and items that they are designing fit the users. These contributes to the following aspect of the final systems being delivered:

- Vision is usually the primary channel for information, yet systems are often so poorly designed that the user is unable to see the work area clearly. Many workers using computers cannot see their screens because of glare or reflections. Others, doing precise assembly tasks, have insufficient lighting and suffer eyestrain and reduced output as a result.
- Sound can be a useful way to provide information, especially for warning signals. However, care must be taken not to overload this sensory channel. A recent airliner had 16 different audio warnings, far too many for a pilot to deal with in an emergency situation. A more sensible approach was to have just a few audio signals to alert the pilot to get information guidance from a visual display.
- Job design is one goal of ergonomics is to design jobs to fit people. This means taking account of differences such as size, strength and ability to handle information for a wide range of users. Then the tasks, the workplace and tools are designed around these differences. The benefits are improved efficiency, quality and job satisfaction. The costs of failure include increased error rates and physical fatigue—or worse.
- Human error in some industries the impact of human errors can be catastrophic. These include the nuclear and chemical industries, rail and sea transport and aviation, including air traffic control.

When disasters occur, the blame is often laid with the operators, pilots or drivers concerned—and labelled 'human error'. Often though, the errors are caused by poor equipment and system design. The contribution of an ergonomist on the management team of a VLITP working in these areas pay particular attention to the mental demands on the operators, designing tasks and equipment to minimise the chances of misreading information or operating the wrong controls, for example.

DESIGNING VLITP BY ERGONOMIC RULES

The purpose of using ergonomics rules in the design process of a VLITP is to ensure varies parts of the systems suits the needs of the end-users. Several systems in use today show how the designers are often far removed from the end users, which makes it vital to adopt an ergonomic, user-centred approach to design, including studying people using information technology, talking to them and asking them to test different stages. This is especially important with 'inclusive design' where everyday business activities are designed with very busy and even disabled users in mind. The number of people in the UK aged 75 and over is forecast to double over the next 50 years (Department of Health, 2001). As such, there is a need to extend the range of application, services and systems designed for the general population.

Governments around the world are taking initiatives to the quality of life for the older generations in our society today. In 1995, the UK Government's Office of Science and Technology initiated 'EQUAL' (**E**xtending **QUA**lity **L**ife). This initiative draws research activities together that focus on achieving a better lifestyle, participating more fully and actively, and avoiding or alleviating the effects of disability. In 1997, the built environment was highlighted as an area in which EPSRC (Engineering and Physical Sciences Research Council) funded research could make significant contribution to the aims of EQUAL. Since then several projects have been funded at to contribute to this worthy cause (www.agenet.ac.uk).

Ergonomics use data available on relevant aspects of the capability of the whole population including older and disabled people. Some of these aspects include:

- Physiological (for instance, range of limb movement, strength, vision, hearing)
- Psychological (for example, cognitive, reaction time, memory)
- Anthropometric data is also required (size and shape ranges of people).

With data such as this available, a knowledge base can be generated for access by conscientious design in VLITPs. The systems should therefore be able to contribute to the 'quality of life' for older and disabled people by enhancing the project to bring improvements in their working environment. This may involve simple things like sensory aspects (acoustics, lighting, comfort, communication systems, signage and navigation).

ERGONOMICS IN THE HEALTHCARE INDUSTRY

This sections pulls together two policy streams—those of patient safety and of older people—and links them with standards in the UK National Health Service. Following a report documenting the extent of medical error in the UK, the government committed to reducing errors (Department ofHealth, 2004). Medication errors were a particular concern, and there was a commitment to reduce serious errors in prescribed drugs by 40% by 2005. Another report by the NHS Chief Pharmaceutical Officer detailed the particular types of errors which can occur in the various healthcare environments, and notes how GP repeat prescribing systems may make it difficult to ensure that therapy is adequately monitored or reviewed. The report contains—the national service framework for Older People—the recognition of the special needs of the elderly adding to the accomplishment of an earlier report that recognised the special medication needs of older people, and the need for regular review (Department of Health, 2001). There are many more barriers (and it requires significantly more resource), to undertake patient-focused research in this setting, than when patients are in hospital or in their own homes. These difficulties relate particularly to recruitment of the homes themselves (as they are largely autonomous), gaining the consent or assent of the elderly and often vulnerable residents, and then in obtaining reliable information from the residents, their carers and medical records (in the home and the surgery).

In the field of patient safety it is likely that medication errors are the greatest cause of harm. It has been estimated that 5% of hospital admissions, an admission rate similar to that of cancer, are a result of avoidable harm from medicines (Hepler and Segal, 2003). Although work is emerging on the types of problems in primary care, there is little work on prevalence; what work there is often suffers from a poor definition of an error, and use of methods such as spontaneous reporting, which miss the vast majority of errors.

The elderly are particularly at risk. They have a high level of morbidity, often with multiple health problems and hence need to take several medicines. In addition frailty, changes in drug distribution, and susceptibility to renal and hepatic impairment all mean that these patients are more susceptible to adverse drug events. Evidence from the UK suggests that 19% of admissions of elderly people to hospital are as a result of therapeutic misadventure (Cannon and Hughes, 1997). According to evidence presented to the House of Commons Select Committee on Elder Abuse on 22 January 2004, less than half (44%) of the care homes currently met or exceeded its national standards for handling and administering medicines (Health Select Committee, 2004). The report acknowledged concerns over prescribing and medicines use, and makes recommendations to improve practice.

There is a wealth of research that illustrates inappropriate prescribing in nursing homes in the UK and USA. Studies conducted in the USA draw attention to many lessons drawn (Baker et al, 2003; Gurwitz et al, 1994). Few have been prospective studies of prescribing or drug administration error. There is also a body of literature expressing concern about prescribing to older people, though some are framed as 'inappropriate' prescribing, rather than prescribing error. Although errors are a sub-set of inappropriateness, very little is known about the extent to which the two literatures overlap. Used alone, medication appropriateness scales can produce many false judgements that prescribing is inappropriate, while missing prescribing errors. Nevertheless, studies of prescribing in UK homes suggest inappropriate prescribing occurs in between 50 and 90% of patients (Alldred et al, 2003).

The objective of ergonomics in VLITP for the healthcare sector would therefore be as follows:

- To establish the prevalence, types and underlying causes of medication errors, and the ensuing harm.
- To develop solutions to reduce the prevalence of error to investigate the prevalence of different types of medication error (prescribing, monitoring, administration, dispensing, interface) with sufficient precision to provide a typology of the errors.
- To understand the underlying causes of errors.
- To apply the principles of ergonomics for:
  a. Analysing existing systems
  b. Understanding and mapping the medication pathway.
  c. Understanding and mapping the information pathways.
  d. Generating, evaluating and prioritising solutions.

In order to assess the causes of error the ergonomist need to establish not only *what* happened but also *how* it happened, before going on to assess *why* it happened.

The question **'what happened'** should be answered by specific research methodologies that include the classification of errors using the following taxonomies:

- *where* they occur in the *medication pathway* (prescribing, monitoring, interface / transcribing, dispensing or administration)
- the *type* of medicine involved (e.g., analgesics, anticoagulants, or neuroleptics)
- the *route* of the medication (e.g. oral, IM injection or suppository)
- the *complexity* of the dosage schedule.

The question **'how it happened'** should be answered using human error theory (Dean and Barber, 1999). Reason extended Rasmussen's Skill, Rule, Knowledge (S-R-K) classification of actions to identify types of errors or mistakes. Skill-based errors are those where the individual is conducting a task in which he or she is skilled, gets interrupted and then makes a mistake. Reason refers to these as slips or lapses. The other two types of mistakes refer to planning or problem-solving failures. Rule-based errors refer to the misapplication of a good rule, e.g. prescribing one drug but writing the dose of another, similar type of drug or similar sounding drug. Knowledge-based errors occur in novel situations, with the use of inaccurate or incomplete information and where the person is subject to biases. Some errors may not be easily classified into one or other error type using this taxonomy by those assessing errors, particularly without knowing exactly what type of action is being executed at the time. This does not only depend on the task, but also the level of experience and expertise of the individual, and the context in which the individuals are working at the time such as the level of interruptions, distractions, etc.

The question **'why it happened'** should be assessed using analysis of risk and safety framework (Vincent et al, 2000). This framework identifies the cause behind the error (i.e. the immediate conditions that enable the errors to occur as well as the underlying or root causes of error). The framework is based on the error theories mentioned above and in particular Reason's Organizational Accident model (Reason, 1997) where both active failures (the errors made by healthcare providers) and latent conditions (the underlying management decisions that affect work conditions) are considered. The framework considers the following factors when assessing the causes and contributory factors of each case: the patient characteristics, including their condition, the individual member of staff, the task itself, the environment in which they work, the team including communication and supervision, and the organizational factors that influence the work environment and

Figure 8.2. Ergonomic requirement for healthcare

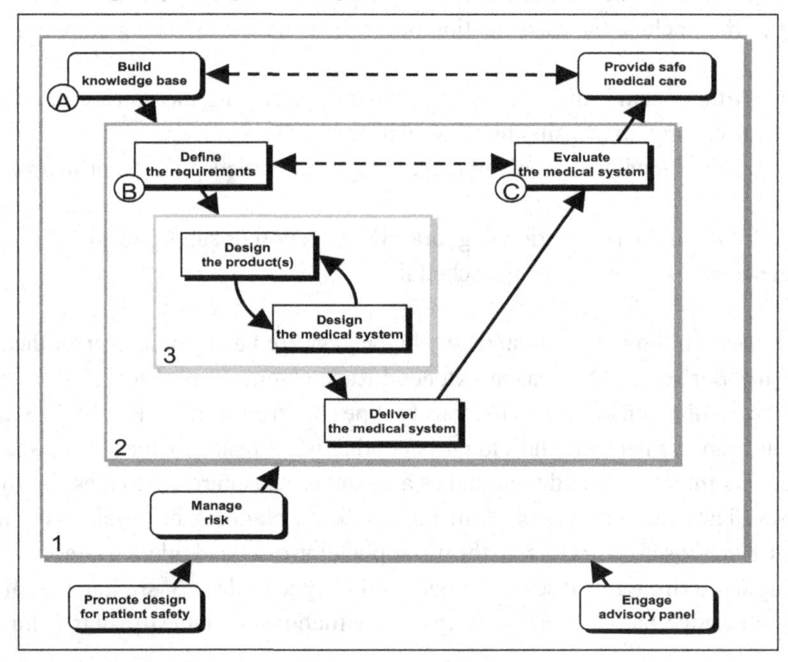

conditions, such as policy and economic decisions. The framework has been used as a tool for the investigation and analysis of critical incidents and adverse events and has been incorporated into a record review methodology (Vincent et al, 2000). As such it can be applied to various methods including interviews, observations and reviewing documents.

An earlier study, published by the Department of Health and the Design Council (Buckle et al, 2003) considered how a system-wide design led approach might help tackle patient safety in the Case Study I (see Chapter XIII). The approach to be taken in this study will take forward the research and design led system advocated in that report. Figure 8.2 shows the essential steps in the process that lead to improved design to reduce medication errors involves building an improved knowledge-based that can then lead to appropriate definition of design requirements.

Whilst it might seem impossible to entirely remove errors in medical systems, it is well established that errors can be minimized by predicting the form that are likely to be taken. The success of such an approach however depends on a thorough understanding of the healthcare system (i.e. A) and how healthcare users and patients really behave within it (i.e. C), this often different to how they are expected to behave.

This involves an understanding of how a range of users interacts with the complex range of medications, equipments, information and each other. The ergonomics approach informs many approaches for improving the performance of complex systems. It is well established in many safety critical industries (e.g. aviation, nuclear, chemical, rail) as an important approach for both evaluating current systems and for developing design criteria that must be met in order to enhance safety. Whilst the ergonomics approach cannot guarantee simple solutions, it's validity in developing not only a better understanding of where and how systems are failing, but also in it's ability to provide a firmer basis on which design requirements can be constructed. These also enable better-informed recommendations and, in some instances, provide the basis upon which designers can advance new and practical suggestions.

CHAPTER SUMMARY

Much of today's human factors research and expertise is channelled towards improving the ways we use information. Virtually everyone has experienced the frustration of using computer software that doesn't work the way they expect it to. For the majority of end users of information systems, if the system is not working they have no recourse but to call for technical help, or find creative ways around system limitations. This is usually in the form of using those parts that are usable, and circumventing the rest or increasing stress levels by using a substandard system. The management team of VLITPs tries to avoid this with a more complete understanding of the users' tasks and requirements had been present from the start. The development of easily usable human-computer interfaces is a major concern for agronomists within very larger projects. This also involves the design of appropriate signs, symbols and instructions so that the end users can quickly and safely understand their meaning.

REFERENCES

Alldred, D. P., Zermansky, A., Petty, D. R., & Raynor, D. K. (2003). Clinical medication review by a pharmacists for older people in care homes: Preliminary report. *International Journal of Pharmacy Practice, 11*, 90.

Barker, K. N., Flynn, E. A., Pepper, G. A., Bates, D. W., & Mikeal, R. L. (2003). Medication errors observed in 36 health care facilities. *Archives of Internal Medicine, 162*, 1897-903.

Buckle, P., Clarkson, P. J., & Coleman, R., (2003). Design for patient safety: A system-wide design-led approach to tackling patient safety in the NHS. London: Design Council/Department of Health.

Cambridge, R.S. (2003). Designing for patient safety. *A scoping study to identify how the effective use of design could help to reduce medical accidents*. Cambridge: Cambridge University Press.

Cannon, J., & Hughes, C. M. (1997). An assessment of the incidence and factors leading to drug-related hospital admissions in the elderly. *European Journal of Hospital Pharmacy, 3*, 14-8.

Dean, B. S., & Barber, N. D. (1999). A validated, reliable method of scoring the severity of medication errors. *American Journal of Health-System Pharmacy, 56*, 57-62.

Department of Health (2001). Medicines and Older People: Implementing medicine-related aspects of the NSF for older people. London: Department of Health.

Department of Health (2004). Building a safer NHS for patients: improving medication safety. London: Department of Health.

Gurwitz, J. H., Sanchez-Cross, M. T., Eckler, M. A., & Matulis, J. (1994). The epidemiology of adverse and unexpected events in the long- term care setting. *Journal of the American Geriatrics Society, 42*, 33-8.

Health Select Committee (2004). An inquiry into Elder Abuse: Evidence from the National Care Standards Commission. http://www.carestandards.org.uk/press+release/ pressitems/elder_abuse.pdf.

Hepler, C. D., & Segal, R. (2003). Preventing medication errors and Improving Drug Therapy Outcomes. A management systems approach. Boca Raton: CRC Press.

Kaber, D.B., Ma, R., Segall, N., Sheik-Nainar, M. A., & Perry, C. M. (2006). Effects of physical workload on cognitive task performance and situation awareness, *Theoretical Issues in Ergonomics Science*, *9*(2), 95–113.

Kim, H-W, & Kwahk, K-Y, (2007). Managing readiness in enterprise systems-driven organizational change, *Behaviour and Information Technology*, *27*(1), 79–87.

Midden, C., & De Vries, P. (2006). Effect of indirect information on system trust and control allocation, *Behaviour and Information Technology*, *27*(1), 17–29.

Mohamed, S., & Irani, Z. (2004) 'Validating Indirect Human Costs' MEFM Taxonomy: Case Studies in Investment Banking' Proceedings of the 10th American Conference on Information Systems.

Neale, W. S., Smerek, R., Haaland, S., Denison, D. R., & Gillespie, M. A. (2007). Linking organizational culture and customer satisfaction: Results from two companies in different industries. *European Journal of Work and* Organizational *Psychology, 17*(1), 112–132.

Reason, J. (1997). *Managing the risks of organisational accidents.* Aldershot: Ashgate.

Stanton, N. A., & Stammers, R. B. (2008). Bartlett and the future of ergonomics *Ergonomics*, *51*(1).

Stappers, P. J., Dekker, S., & Flach, J. M. (2007). Playing twenty questions with nature (the surprise version): reflections on the dynamics of experience. *Theoretical Issues in Ergonomics Science*, 9(2), 125–154.

Vincent, C., Taylor-Adams, S. E., Chapman, E. J., Hewett, D., Prior, S., & Strange, P. (2000). How to investigate and analyse clinical incidents: Clinical Risk Unit and Association of Litigation and Risk Management Protocol, *British Medical Journal*, *320*, 777-81.

Chapter IX
Service-Oriented Architecture:
A New Platform for Very Large IT Projects

ABSTRACT

For centuries, organizations have been trying to exchange information between their applications by linking them together. However, such application integration has not been as successful as organizations have hoped. With the introduction of SOA, application integration is more successful than the previous integration techniques. SOA is a design philosophy in which resources are cleanly partitioned into remotely accessible software components performing self-contained functionalities, called services. The reinvention of SOA in recent times is attributed to the rise of Web Services, which has become commonly used in VLITP to expose services within the host organization. However SOA can also be implemented with other service exposing techniques. SOA is based on the concept of separation of concerns, realizing that no single entity can be best at everything. SOA is usually implemented using an Enterprise Service Bus (ESB). The ESB is responsible for routing, prioritizing, scheduling, monitoring, and controlling the flow of traffic between services and therefore forms the middleware for Service Orientation.

INTRODUCTION

Service Oriented Architecture (SOA) is a design philosophy describing an architecture in which the applications expose VLITP functionalities as services. These services are remotely accessible software components that perform specific tasks. The organization's applications can be integrated by assembling services. This can be considered integration, because assembling services is possible regardless of their location. The combined services provide support to or are responsible for the execution of a business activity. Service Orientation has become popular, because it increases the modifiability of the applications by separating the process flow and the exposed functionalities.

SOA has elements that help reaching its full potential. The open standard based web application technology such as Web Services is necessary to give SOA the flexibility it is known for. Web Services are gaining more popularity yearly, especially because they are based on the well known easy to use XML language (Gottschalk, 2002). The many advantages these Web Services provide made them the most used open standard in combination with SOA.

Another important element of a SOA is the Enterprise Service Bus (ESB). A SOA environment needs to have a component which takes responsibility for delivering service requests. For this task, an ESB is mostly used, especially because it was built for operating in a complex infrastructure consisting of many different applications. Although Web Services and ESB are important in a SOA, without good governing SOA is doomed to fail. A SOA needs to be managed and controlled. Therefore, SOA governance is another vital element in a service-oriented environment.

CONCEPT OF SOA

SOA is a framework that offers application integration based on the concept of providing independent or loosely coupled services. It is viewed from a technological perspective as well as a business perspective not just as a result of previous integration solutions being too technological driven but also due to its focus on the business requirements. SOA can be expressed as a set of flexible services and processes that a business wants to expose to its customers, partners or internally to other parts of the organization (Bierberstein, 2005). SOA defines software in terms of discrete services, which are implemented using components that can be called upon to perform a specified operation for a specific business task (Bierberstein, 2005).

During the late 1990s many organizations invested heavily in Information Technology. Their ever-growing architectures had become increasingly difficult to integrate and new company-wide applications called Enterprise Resource Planning (ERP) systems emerged. These systems had been designed to integrate and optimize various business processes such as order entry and production planning across the entire organization (Mabert et al, 2001). For example a clerk in the delivery department registered that new goods have arrived. The ERP software automatically triggered an update of the stock ledger and adjusted the machines' production schedules.

ERP packages were advertised as end-to-end solutions, which could replace all other existing application. However many organizations discovered that these packages could not provide all the required functionalities and had to implement ERP within the existing architecture (Mabert et al, 2001). ERP package had become one of many applications in organizations. Some organizations even use multiple ERP packages.

The biggest disadvantage of implementing a single ERP package is that it only supports a limited amount of implementations of a business process. Organizations have to redesign business processes to fit the software. Although it is possible to customize ERP manually, many project leaders have had the experience that "going to war" with ERP leads to overrunning costs and total project failures (Summer, 2000). Many managers avoid these risks and therefore choose for short time profits by adapting the business process to the application (Guah and Currie, 2005). They have been defending this strategy by pointing out that ERP vendors have much more experience with good management of processes, because of their involvement in many organizations in multiple industries. The software is therefore correct and the organization's process has to be redesigned. Thus instead of adapting the software to the business needs, many organizations have been adjusting their process to fit software. However, redesigning processes to fit software is highly undesirable, because:

- The redesigned process is not the best process implementation. The organization would have chosen a different implementation if there was an application suiting their needs. Therefore the redesigned process will perform sub optimal at best.
- Many employees have a natural aversion for changes. Changing to undesired situation will only strengthen this resistance.
- If competitors change their processes similar to yours then you will loose your competitive advantage. It is even possible that the only market strategy becomes Operational Excellence (Treacy and Wiersema, 1993).

ERP implementations can improve the efficiency of the related business processes. However, these processes can be improved even more with tailor made applications. The broad variety of desired functionalities makes it clear that the software industry will never be able to build a single application fulfilling all these desires. Consequently the industry's attempt to take away the integration desire by making ERP packages has failed. Although that tailor made applications can improve the business process to an optimum level, they are more expensive to build and implement. This is why the software industry has returned to software integration of smaller applications.

The high failure rate of integration projects, which was as high as 70 percent in 2003, made organizations reluctant to start new expensive integration projects (Trotta, 2003). The software industry has to standardize integration to reduce the failure rate. One of much discussed new integration approaches is the SOA. This Service Orientation paradigm is based on exposing functionalities of any number of applications, called services, to other application through a platform independent interface. The goals that are pursued with Service Orientation are identical to the goals of any other application integration technique. These goals consist of improving the IT support to the business process. Several developers feel confident that Service Orientation enables them to successfully integrate organization's current architectures and to set a basis for future integration projects (Boeré, 2006; Cordial, 2006).

Figure 9.1. Decoupling applications through services (IBM, 2006)

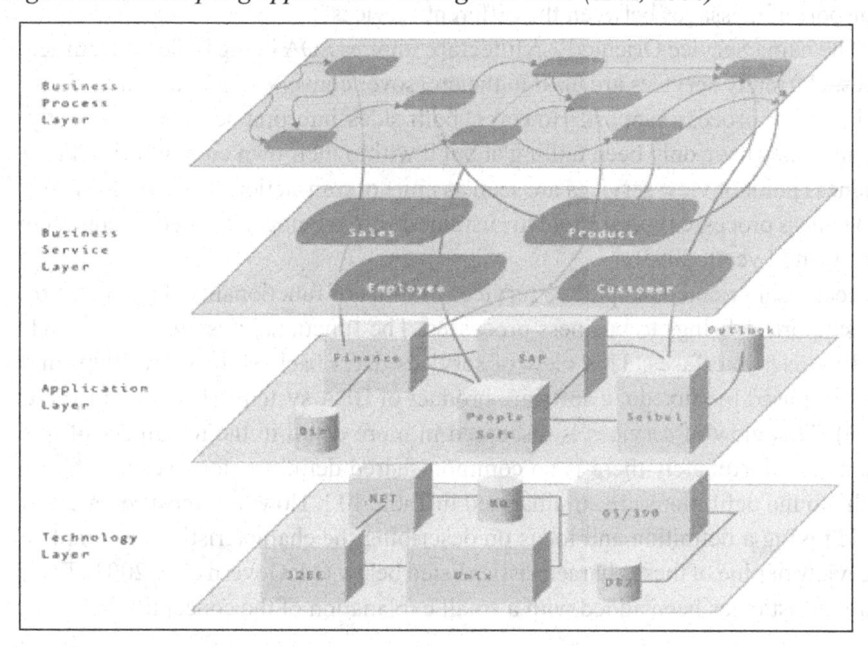

Services can be described as a group of components that perform discrete functions, which operates independent of each other and are invoked by interfaces. These services, on the other hand, can be combined to form composite applications that deliver higher order business functionality (TIBCO, 2006). Figure 9.1 illustrates how applications, within a SOA, can be divided into reusable "services" containing four layers:

- **Technology layer:** A layer seen as the core layer for platform integration.
- **Application layer:** Where applications from different vendors are connected with each other.
- **Business service layer:** Where the services get separated from the lower levels.
- **Business process layer:** Through which messages get invoked to reach the interfaces for services in the business process layer

A vital element of the SOA enabling technology is Web Services because they are open standard based Web applications that allow application integration. The open standard use of the "Extensible Markup Language" (XML), which is an operating system and language independent, can implement the "services" in a SOA using Web Services. Another important element in the SOA environment is the "Enterprise Service Bus" (ESB). This is a piece of middleware that manages the transport of messages between the different services.

The name Service Oriented Architecture implies SOA being build up from services. Currently services are used in the crossover conversation between business and IT over process support. However, both sides interpret services differently, because they have only been talking about it within their own communities. From business point of view services are seen as units of transaction. They are the result of business processes and are sold to customers, often being described by contracts (Perry and Lycett, 2003).

Technical oriented people see services as units of functionality. They focus on the support IT brings to business processes. The functionalities are often visualized by user interfaces. This view of services dates back to the early 1990s and was first used by Tuxedo, a software product of BEA systems (Perry and Lycett, 2003). This view of services is discussed in more detail in the remainder of this chapter. Unfortunately there is no common shared definition for a service. Some of the found definitions are summarized in Table 10.1. However, most researchers avoid giving a definition and focus on describing the characteristics. A complete overview of nine of these characteristics listed below (McGovern et al, 2003). Each characteristic has been added with a small explanation of the concept:

- **Services are discoverable and dynamically bound.** The location of each service is stored in a central registry. This registry will provide the location of the service before it is invoked.
- **Services are self-contained and modular.** The application that contains services is decomposed into smaller modules. Each module or service resembles a specific functionality of the application and should make sense to designers without having knowledge of other services.
- **Services are interoperable.** The service must be able to accept invoke requests from any platform and programming language. The invoke request can therefore not consist of any platform or language specific parameters.
- **Services are loosely coupled.** This characteristic has been mentioned in all articles discussing SOA and is considered as the most important characteristic of a service. Coupling refers to the level of dependency between services. Service A is dependent of B if A requires the functionality provided by B to function properly. Loose coupling refers to a low level of dependency, whilst tight coupling refers to many dependencies. Loose coupling is achieved by a service's discoverability and the ability to define multiple invoke request formats for a single service and is desirable because it increases the modifiability of the architecture.
- **Services have a network-addressable interface.** Each service is exposed to other applications in the Service Oriented Architecture. The applications can call the service with a service request, which is sent across the network. This could be within a single organization or inter-organizational.
- **Services have coarse-grained interfaces.** The concept of granularity refers to the level of functionalities provided by a service. A coarse-grained interface provides a wider range of functionalities than a fine-grained interface. For example a service validating any kind of payment is more coarse-grained than one that can only validate credit card payments (McGovern, 2003).
- **Services are location transparent.** This characteristic is also related to services being loosely coupled. The location of a service is not hard coded into any application, because it can be found in the service registry.
- **Services are composable.** The modular design of services allows services to be built on top of each other. Composing a new system based on combining several existing services is preferable, because these services have already been designed and tested. Reuse of services is also promoted due to the coarse-grained interfaces.
- **Services are self-healing.** An application can crash during the execution of a service. This could have undesirable consequences. For example a banking system could crash during a money transfer. The crash occurred when the money was withdrawn from one account, but not added to another. The self-

healing transfer service will perform a rollback when it is restored from the crash.

The common perspective from the definitions given in Table 9.1 is that a service is a software component representing a self-contained functionality. These components also have to comply with the nine characteristics mentioned earlier. Unfortunately it is hard the judge if a component meets the demands regarding the concepts of granularity and coupling, because these concepts are subjective. For example there is no clear line that separates a fine-grained and a coarse-grained interface, because one can always define a more coarse-grained interface than the one it is currently using. Whether a certain interface can be considered coarse-grained therefore depends on the scope of the application and the opinion of the architect.

Services are frequently described as fundamental building blocks like Lego. One can build several complicated structures by combining or assembling these building blocks in a different order. This flexibility also applies to services, because they can be assembled by sequentially invoking them. Assembling services is only

Table 9.1. Definitions and technical interpretation of a service

Definition	Reference
Services are network addressable entities with a well defined, easy-to-use and standardized interface.	(Bocchi and Ciancarini, 2006)
A service represents some functionality (application function, business transaction, system service, etc.) exposed as a component for a business process.	(Dodani, 2004)
A service is a unit of work done by a service provider to achieve desired end results for a service consumer. Both provider and consumer are roles played by organizational units as well as software agents on behalf of their owners.	(Hao, 2003)
A service is a remotely accessible, self-contained application module.	(Krafzig et al, 2004)
Each system in a Service Oriented Architecture exports own features to the network as a unit of service (a set of tasks, which is coarser than an object).	(Nakamura et al, 2004)
A service is a contractually defined behavior that can be implemented and provided by a component for use by another component.	(Guah and Currie, 2005)
Services are self-describing, platform-agnostic computational elements that support rapid, low-cost composition of distributed applications.	(Papazolgou, 2003)
A service is a software component that is described by meta-data, which can be understood by a program. The metadata is published to enable reuse of the service by components that may be remote from it and that need no knowledge of the service implementation beyond its published meta-data.	(Schmidt et al, 2005)
Similarly to objects and components, a service is a fundamental building block that - Combines information and behaviour; - Hides the internal workings from outside intrusion; - Presents a relatively simple interface to the rest of the organism.	(Sprott and Wilkes, 2004)

possible in an architecture consisting of multiple services. None of the found SOA definitions in literature discusses the number of interacting applications. In theory it is therefore possible that SOA consists of one application exposing a single service. However in general it will consist of a large number of services spread over various applications. The extra abstraction layer introduced by Service Orientation is expensive to build, because it usually involves modifying applications to enable exposure. These costs can only be justified by simplifying a complex architecture or by enabling integration between multiple applications. Next to cost justification, it does not make any sense to decompose a single application into services and to rebuild it providing the exact same functionalities. This is why a service has to operate in an architecture consisting of multiple other services spread over various applications.

There is also no commonly shared definition for SOA. SOA is not an application, neither is it Web Service, nor an out of the box solution to solve integration problems. Instead, SOA is a software paradigm based on exposing functionalities of any number of applications—called services—to other applications through a platform independent interface. It therefore adds an extra layer to the levels of coding that can help programmers to find the required functionalities without getting tangled up in the implementation of other applications. Whether this solves integration problems still depends on the skills of the programmers.

WHY SOA?

In the 21st century, characterised by a competitive and constantly changing business environment, the agility of an organization is of paramount importance. Flexibility needs to be increased in order to react much faster to new market demands and to the opportunities that present themselves. Regarding network infrastructure, integrating business applications in an easy, less time consuming, inexpensive and effective way is the key to this flexibility. In instances where an organisation purchases many different systems and applications from more than one vendor, this becomes a challenge, especially when these systems and applications are legacy in character. Many organizations rely on these rather old systems usually because of the unreasonable costs required to replace them.

The vertical and horizontal integration enterprises component has been connecting applications with each other for several decades as a way to integrate systems. Then, this kind of application linking was sufficient. Nowadays, however customers are demanding faster and better services which mean productivity needs to be increased. At the same time shareholders are pressuring to lower costs. In order to compete in this fierce environment faster anticipation to changes are required.

Figure 9.2. Evolution of business integration

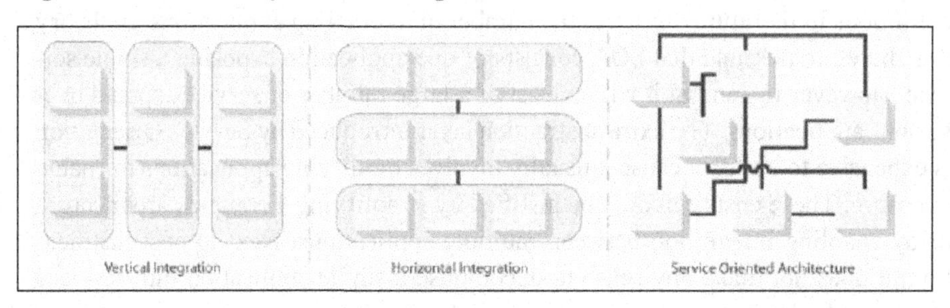

New integration techniques are needed to offer flexible business applications. SOA offers this opportunity.

Figure 9.2 shows the evolution of business integration. In the 1980s and earlier applications and systems were integrated vertically. This implies that applications and systems of a value chain were connected with each other. In the 1980s and 1990s integration between systems and applications was expanded across vertical lines, making horizontal integration possible. Here, applications and systems at the same level of the value chain were connected with each other. Nowadays SOA is the current type of integration. In a SOA, applications and systems at different levels of the value chain are integrated with each other.

For the current business environment, the implementation of SOA using Web Services is considered the "state of the art" in systems integration (IBM, 2006). SOA provides the flexibility needed by creating architecture that makes systems and applications operate independent from each other. This independency increases the speed at which products and processes are implemented or modified. As a result the enterprise can meet market demands much faster than before.

With open standards such as Web Services connections between systems no longer need to be hard coded (Zhang, 2005). This custom coding is not considered efficient because as it only applies to a particular application. This means that the coding cannot be reused. One change in the code usually requires changes in other applications as well. This is not only time consuming, but also expensive. SOA in combination with Web Services makes it possible to integrate much easier and faster, additionally the services can be reused. This gives SOA a big advantage compared to other integration technologies.

SOA Benefits

Considering that organizations need to be flexible in order to survive in this economy and given that flexibility is an important goal of SOA, implementing VLITPs via

SOA provides the host organization with many opportunities for IT as well as new businesses. Literature research on the benefits of SOA shows that a SOA has many rewards. The major benefits of SOA are as follows:

- **Reuse of IT services:** In a service-oriented environment, systems and applications are divided into independent reusable services. Independent means that the services are not developed for a particular system or application, but can be easily used for different systems or applications. This independency makes the services reusable.
- **Cost reduction***:* By reusing services, not many new services will have to be development. As a result, development costs will decrease. Furthermore, services can be easily modified, because they do not rely on a particular system or application. This means that a modification of a service does not require many modifications on services that integrate with the modified service. As a result maintenance costs will reduce as well.
- **Speed to market:** By reusing services, less time will be spent designing and building new ones. With this reusability, enterprises can offer products much faster to the market. At the same time, the easy modifications of services create more flexibility. This flexibility gives organizations agility. As a result, organizations can adapt faster to new market demands.
- **Better integrating business with IT***:* Nowadays, business is heavily depending on IT. Connecting business and IT while IT aligns with business requirements remains a difficult task. SOA integrates business and IT into a framework that simultaneously leverages existing systems and enables business change (Feeny, Lacity and Willcocks, 2006).
- **Better and faster decision-making:** Within a SOA environment, IT is better aligned with business requirements. As a result, information within the organization is shared and understood better than before. This allows an organization to make better and faster decisions.

While the identification of many benefits to a service-oriented environment is cleared to see, it is not certain if there are any disadvantages. Such issues will be the focus for the following section.

SOA Disadvantages

While above listed advantages can prove to be impressive to organizations, there few disadvantages which are not common in the IT literature—partly because of current hype of enjoyed by SOA at the moment. The impressive seems to be that SOA is that integration resource organizations have been waiting for such a long time. However, moving to a service-oriented environment does not come without its difficulties. Only three of the major once have been mentioned below:

- Although SOA might increase control complexity, going for a SOA means dividing applications and systems into independent services. Instead of controlling one application, numerous independent services from that application need to be controlled, thereby presenting possible challenges to be managed. A good control framework at this stage seems inevitable.
- The concept of SOA is easier to explain than to implement because implementation tends to be a difficult task. Thus, SOA experts are warning organizations about implementation failures.
- Finally, the benefits from SOA—mentioned above—would probably not emerge within a short time period. As a result of many SOA benefits being long-term rewards, most implementations of SOA do not appear as successful projects.

THE ARCHITECTURE

Most SOA have a client-server communication infrastructure which consists of three components:

1. The service consumer
2. The service provider
3. The service registry

A service consumer is a client or simply a consumer, who wants to obtain a service. A consumer can either be seen as an application, some other type of software module or just as another service.

A service provider is a server, which provides a service. The term "server" should not be seen as a computer (server), but as a component or some type of software system. A service provider can also be a service consumer at the same time, as it occasionally needs to acquire another service in order to meet the consumer's request. A service registry is a directory that records available services. It does not actually store the service, but it collects information about the service and its relation to other services.

Together the afore mentioned three components interact with each other by using the operations "publish", "find" and "bind". "Publishing" a service simply implies notifying the service registry of a (new) service. Service providers need to make their service publicly available by providing their location and other relevant

Figure 9.3. SOA components and operations

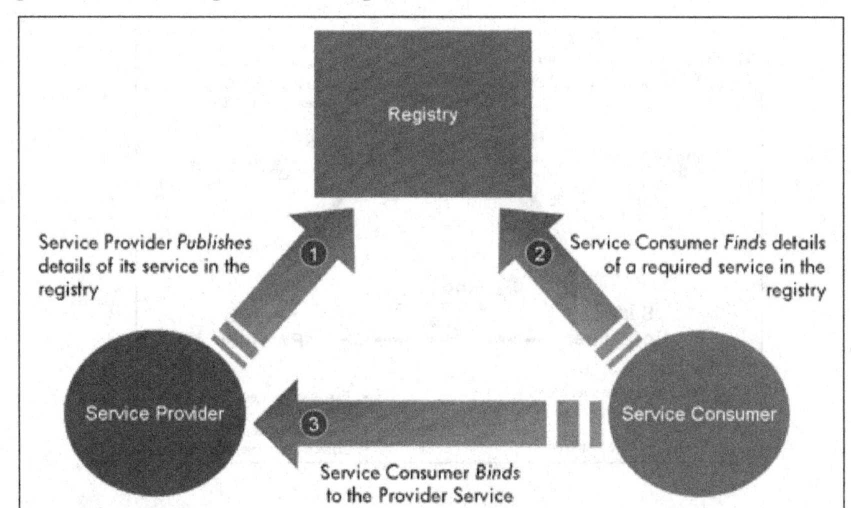

information to a registry. The "find" operation is used to find or locate a service. This can be done by browsing into the registry for gathering data which can tell a service consumer where and how to acquire a particular service. The "bind" operation connects the service consumer with the service provider. By binding these together the service can be provided. The interaction between the components is illustrated in Figure 9.3.

From Web Services to SOA

The Internet made information exchange easier, faster and globally possible. However it was not always possible for applications to communicate with each other, due to incompatibility (Guah and Currie, 2005). Web Services however have the potential to overcome this boundary, because it is based on open standards like XML, URL and HTTP which are widely accepted. Open standards provide a way to operate independent from languages, formats and platforms. Figure 9.4 illustrates how Web Services standards are used—with many of these models including the standards mentioned previously in this chapter.

According to the presented model, WSDL and UDDI are used for publishing and finding a service, while SOAP is used for binding of a service. In the next section the model is explained.

The interaction as presented in Figure 9.4 is described as follows:

Figure 9.4. Interaction of Web Services in a SOA

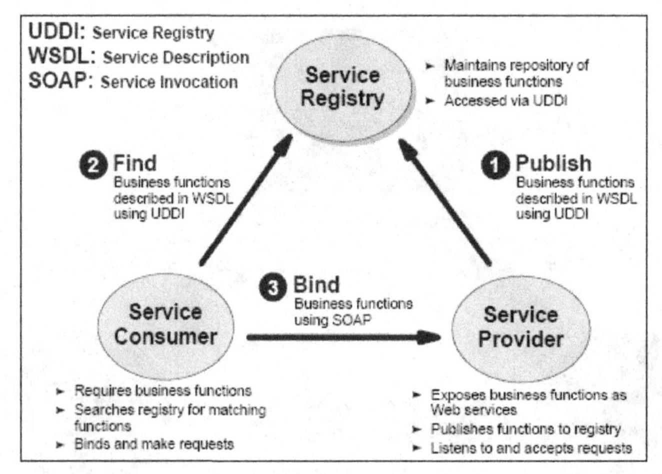

1. A service provider can offer business functions as Web Services. The services' interface and access information are described in WSDL and are made publicly available by publishing them to UDDI, the service registry. The UDDI now contains the information of these Web Services.
2. The service consumer that wants to use a service from the service provider has to browse through the service registry to find the WSDL service information.
3. When the location of the service is found, a request is made to the service provider in order to invoke the service. The invoking is done in a format according to SOAP.

Enterprise Service Bus

A key element of a SOA is the ESB, which manages the communication between the services. It can be outlined as a piece of middleware that connect applications together. It is usually illustrated as a pipe than runs through a company's IT environment (see Figure 9.5). As a result, people think that an ESB is a component or tool. In reality, it is a set of capabilities that provide end-to-end connections between the different IT resources.

Communication through a bus connection reduces complexity, which is considered a common problem within an enterprise. Integration through point-to-point connections can often become very complex when a company grows. As a result they acquire more software. The applications are all directly connected with each

Figure 9.5. Reducing connection complexity with an ESB

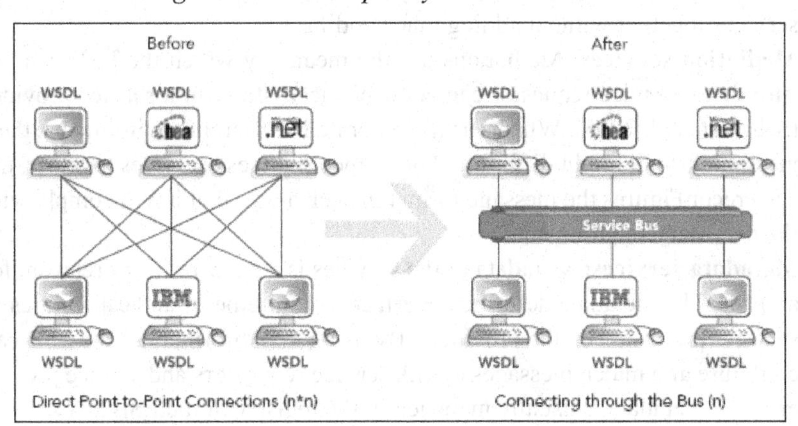

other, which result in a 'spider web like' image. Applications are therefore more difficult to manage. Using an Enterprise Service Bus, the software components are integrated by only connecting to the bus. The bus will then further establish the link between the components.

Although the ESB is usually seen as a SOA component, it can also operate in a non- Service-Oriented environment. It should also be mentioned that a SOA could be built without an ESB. However, an additional tool that manages the connections of the services is then required.

The ESB is not only a connection enabler. One of the additional values of the ESB is that it covers the responsibility for many of the infrastructure concerns that might otherwise surface in application code. The capabilities the ESB offers are:

- **Messages services:** The messages (also known as service requests) which are sent between the services are managed by the ESB. It does not matter what type of message is used, as the ESB can handle any type. Furthermore the ESB guarantees that messages are delivered and can also be divided or combined.
- **Management services:** For an ESB it is necessary to have a self-management system to monitor its own performance and to correct when a message delivery failed. When traffic flow is high, interference grows. In order to make sure that all the tasks are performed correctly, the ESB needs to notice the failure and applies corrections where necessary.
- **Interface services:** In order for services to connect to an ESB, they need to have an interface. Even thought these interfaces can greatly differ from each other, the services still need to be connected. Therefore, the ESB has to be

able to support and provide many types of interfaces in order to connect the services together without adding much coding.

- **Mediation services:** Mediations are the means by which the ESB can ensure that a service requestor can connect successfully to a service provider (Schmidt et al, 2005). When services operate in different environments their message format might differ too. For connecting these services together, the ESB reconFigures the message format in such a way that it will comply with the necessary requirements.
- **Metadata services:** Metadata of the services is considered very relevant for an ESB. The metadata describes interfaces, requirements and capabilities of services published by the providers. By using this information, the ESB can configure and match messages to link service requestors and service providers. The metadata is usually managed in the registry of the ESB.
- **Security services:** Security remains an important issue when information is exchanged over the Internet. Since the Enterprise Service Bus is responsible for connections between services in which information exchange occurs, it is required to deal with security. The ESB offers a framework to deploy security such as integrity, privacy, and authentication.

SOA GOVERNANCE

A service-oriented environment can provide benefits such as cost reduction, better IT alignment with business requirements and faster time to the market. These benefits require a well implemented and managed SOA. As a result, governing the shift to SOA and managing this new environment becomes essential. Therefore, SOA governance is an important element for a successful SOA.

SOA governance is about managing the quality, consistency, predictability, change and interdependencies of services (www.looselycoupled.com). Services are an important part of the SOA environment and need to be controlled. There need to be policies and procedures for all services including the metadata used by these services. To make sure that this architecture is and stays reliable SOA governance is necessary. Moreover, SOA governance can be seen as IT governance in an organization with service architecture. In other words, SOA governance is about controlling IT in a SOA environment.

The main goal for developing a SOA is to achieve a loosely coupled, yet integrated architecture. It consists of three kinds of participants: service providers, service clients, and a service registry. These participants are applications or so called software agents. A service provider is an application that exposes its services to the network. It constantly waits for an invoke request to perform the requested service. The provider registers a service in the service registry by sending a service contract. The registry stores the contract, which contains essential information for invoke requests: a description of the functionality provided by the service, its location, and which security protocols are being used. Each service can be registered by multiple contracts to enable different interfaces for specific users.

A service client or service consumer is an application that requests a service from a provider. The applications are not restricted to a single role, because a provider can also be a client while performing its service. The following scenario describes how the three participants interact. A client is performing a certain business activity, which requires invoking a service. It queries the registry to find a service that suits its needs. The registry responds by sending a service contract to the client. The contract enables the client to bind to the provider and to invoke the service. Depending on the nature of the service the provider can send a reply containing the results. The whole brokering process is frequently illustrated like in Figure 9.6.

Figure 9.6 shows the main concept of Service Orientation in which each provider publishes its services in the registry. The client queries the registry for a required service and receives the information to bind to the appropriate provider. The client's query can differ depending on which type of registry is used. In an ideal situation an automated registry only requires keywords to find the desired functionality (Mukhi et al, 2004). However, most registries have to be queried using the name of the service, which is provided by the developer of the application. The dynamic binding that is provided by the registry results in a loosely coupling between the applications, because it enables developers building software that automatically generates the service request from the service contract that is provided by the registry. Therefore dynamic binding increases the modifiability of the system.

Many early adopters of SOA have not implemented a registry (Schmelzer, 2004). It is questionable if these architectures can be labeled Service Oriented, because without a registry there is no dynamic binding. The reason why the early adopters have not used a registry is that dynamic binding adds extra query and binding steps

Figure 9.6. Service orientation concept

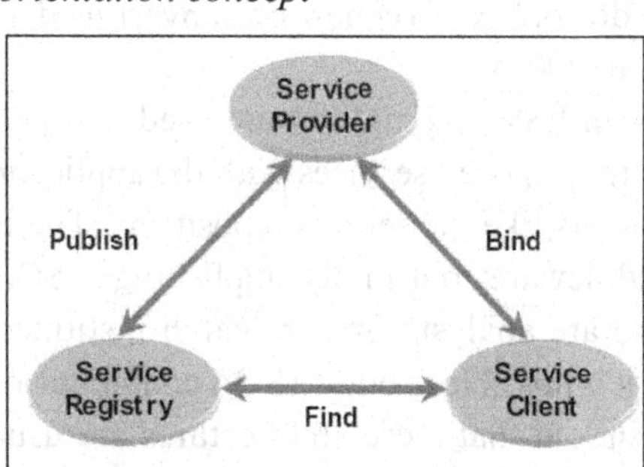

to a business activity. An application not using dynamic binding therefore performs faster than an application that does. It is tempting to code a developer's knowledge about the current location of a service and its service contract. This is especially true when facing high volumes of a certain service. The architect of the system has to choose between a system's performance and its changeability.

SOA is based on the separation of concerns. The consumer's only concern is that the provider performs the requested service. It does not have to be concerned with how a service is executed. This makes it comparable with a Remote Procedure Call. Similar to a brokering process, sending a message from the client to the server initializes Remote Procedure Calls. The server shields its internal workings from the client and can send a response message back to the initializing application. The characteristics that distinguish a service from a Remote Procedure Call are the platform independent request message protocol and dynamic binding. Services usually have messages formatted according to the Simple Object Access Protocol (SOAP). This is an XML based style that has become popular due to Web Services, but in essence services can be implemented using any other protocol.

Enterprise Service Bus

The three participants in SOA are connected via a network. As discussed earlier, middleware has often been used to make it easier to communicate between multiple applications. The Enterprise Service Bus (ESB) is often described as the next generation middleware tool and is illustrated by drawing a horizontal bar with several vertical branches. Each branch depicts a set of services that is provided by an application that can be plugged in or out of the network at any time (see Figure 9.7). The ESB is responsible for routing, prioritizing, scheduling, monitoring, and controlling the flow of traffic between services (Varhol, 2006). Most ESB vendors have also incorporated mediation into their products. Adapters that convert data and platform specific messages to other formats provide mediation. Mediation therefore enables organizations to link non-SOA compatible applications to the ESB. Another advantage of mediation is an increased level of separation of concerns, because low-leveled applications do not have to cope with converting data before being able to process it (Herault et al, 2005).

Figure 9.7 represents an ESB, the middleware used in a Service Oriented Architecture and connects the exposed services with the applications. The ESB also consists of extra components like the service repository. The ESB has long been seen as an essential middleware tool in accomplishing a SOA implementation. However, currently there are analysts from research institutes like Gartner and Forrester that admit that it is possible to build a SOA compliant architecture without using an ESB. They point out that these architectures are usually small and not

Figure 9.7. Enterprise service bus (adapted from Dodani, 2004)

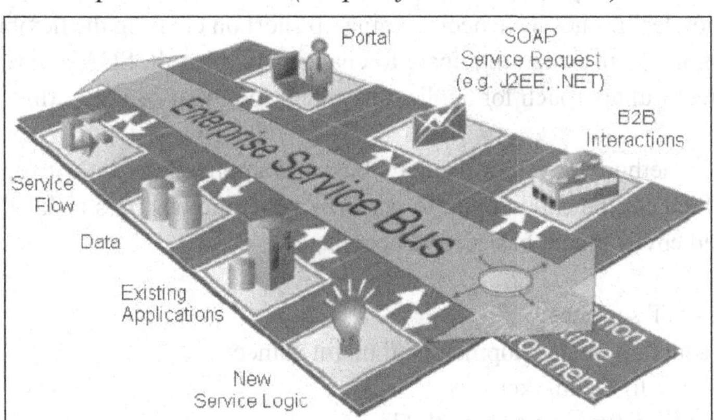

very complicated like test environments. We conclude that the view has changed to ESB being useful, but not required (Chappell, 2006).

SOA in Merger Situation

There was an apparently major integration issue at KLM during the merger with Air-France. There were two IT systems from two different companies that had to be integrated. In order to solve this problem, Air France KLM engaged TIBCO who specialize in the aviation industry. (TIBCO, 2008)

The different IT-components were put into one SOA framework. The goal was to integrate the existing applications of Air France and KLM, to achieve smoother business processes resulting to better efficiency. TIBCO offers Air France and KLM a Complex Event Processing (CEP) software package for real time analysis. This allowed the company to have a clear view of the operational performances and customer needs. By so doing, KLM was more flexible in taking advantage of the changes then taking place in the market. The system allowed KLM to improve its competitive position.

CHAPTER SUMMARY

Organizations operate in an environment that is competitive, demanding and constantly changing. To survive they need to be agile. This agility has to come from flexible business processes. To create this flexibility applications and systems that support these business processes need to integrate smoothly with each other. Thus far integration techniques have been coming up short on creating the flexibility an organization needs to faster anticipate to change. As a result, SOA is introduced as an architectural approach for application integration that can offer this kind of flexibility.

SOA is a methodology that changes business applications into independent services, which can be called upon to perform a specific business task. This service-oriented environment has several benefits:

- Reuse of IT services;
- Cost reduction on development and maintenance;
- Faster time to the market;
- Better integrating business with IT;
- Better and faster decision-making.

Even though SOA offers many benefits, organizations should realize that implementing a SOA is a difficult task. Moreover, a SOA can increase control complexity and the benefits will only be rewarded if the implementation was successful.

This chapter has looked more closely at the architecture itself, and demonstrated a client-server communication infrastructure, consisting of the service consumer, the service provider, and the service registry. These components use the operations "publish", "bind", and "find" to interact with each other.

The Web Services technology together with an ESB makes a SOA real. Web Services are an open standard based web applications that uses XML, URL and HTTP, which are widely accepted standards. With open standards it becomes possible to operate languages and platform independent.

Another key element of a SOA is the ESB—a piece of middleware that connects the services in a SOA together (Bernstein, 1996). In order for a SOA to be successful, it needs to be managed and controlled. Thus, SOA governance is an essential part of its environment. Loose coupling is a very important aspect in Service Orientation. It is mainly achieved by service brokerage, which is the name of the process invoking services. Service brokerage enables developers to build software that is able to automatically invoke service based on the received service contract. This increases the modifiability of the system for both the implementation of a business process as for changed implementations of existing services. However dynamic binding that comes with service brokerage is a heavy burden on the system's performance. This is why many early adopters choose to leave out the registry out of their architecture.

REFERENCES

Bierberstein, N., *Service-Oriented Architecture Compass: Business Value, Planning andEnterprise Roadmap*, IBM, 2005.

Bernstein, P. A. (1996). Middleware: A Model for Distributed System Services. *Communications of the ACM, 39*(2), 86-98.

Bocchi, L., & Ciancarini, P. (2006). On the Impact of Formal Methods in the SOA. *Electronic Notes in Theoretical Computer Science, 160*, 113-126.

Chappell, D. (2006). Trends in the Evolution of the ESB. Retrieved April 2nd, 2007, from http://www.ftponline.com/special/esb2/dchappell/

Dodani, M. H. (2004a). From Objects to Services: A Journey in Search of Component Reuse Nirvana. *Journal of Object Technology, 3*(8), 49-54.

(Feeny, Lacity and Willcocks, 2006).

Gottschalk, K., Graham, S., Kreger, H., & Snell, J. (2002) Introduction to Web Services Architecture. *IBM Systems Journal*, *41*(2).

Guah, M. W., & Currie, W. L. (2005). Web Services in national healthcare: The impact of public and private collaboration. *International Journal of Technology and Human Interaction*, 1(2), 48-61.

Hao, H. (2003). What is Service-Oriented Architecture. Retrieved September 24th, 2006, from http://webservices.xml.com/pub/a/ws/2003/09/30/soa.html

Hérault, C., Thomas, G., & Lalanda, P. (2005). Mediation and Enterprise Service Bus: A position paper. *Proceedings of the First International Workshop on Mediation in Semantic Web Services*, 67-80.

IBM, *Powering SOA with IBM Data Servers*, 2006.

Krafzig, D., Banke, K., & Slama, D. (2004). *Enterprise SOA: Service-Oriented Architecture Best Practices*: Prentice Hall PTR.

Mabert, V. A., Soni, A., & Venkataramanan, M. A. (2001). Enterprise Resource Planning: Common Myths Versus Evolving Reality. *Business Horizons, May-June*, 69-76.

McGovern, J., Tyagi, S., Stevens, M. E., & Mathhew, S. (2003). *Service-Oriented Architecture*. San Francisco: Morgen Kaufmann Publishers.

Mukhi, N. K., Konuru, R., and Curbera, F. (2004, 2004). Cooperative middleware specialization for service oriented architectures. *Paper presented at the 13th Inter-*

national World Wide Web conference on Alternate track papers and posters, New York, NY, USA.

Nakamura, M., Igaki, H., & Matsumoto, K.-i. (2004). Implementing integrated services of networked home appliances using service oriented architecture. *Paper presented at the 2nd international conference on Service oriented computing*, New York, NY, USA.

Papazolgou, M. P. (2003). Service-Oriented Computing: Concepts, Characteristics and Directions. *Paper presented at the Fourth International Conference on Web Information Systems Engineering,* Washington, DC, IEEE Computer Society.

Perrey, R., & Lycett, M. (2003). Service-Oriented Architecture. *Paper presented at the Symposium on Applications and the Internet Workshops (SAINT).*

Schmelzer, R. (2004). How to think loosely coupled. Retrieved January 11th, 2006, from http://www.zapthink.com/report.html?id=ZAPFLASH-05282004

Schmidt, M. T., Hutchison, B., Lambros, P., & Phippen, R. (2005). The Enterprise Service Bus: Making service-oriented architecture real. *IBM Systems Journal, 44*(4), 781-797.

Sprott, D., & Wilkes, L. (2004). Understanding Service-Oriented Architecture. Retrieved September 14th, 2006, from http://msdn.microsoft.com/library/default.asp?url=/library/en-us/dnmaj/html/aj1soa.asp

Sumner, M. (2000). Risk factors in enterprise-wide/ERP projects. *Journal of Information technology, 15*, 317-327.

TIBCO (2006). *Extending the Benefits of SOA beyond the Enterprise*, 2006.

Treacy, M., & Wiersema, F. (1993). Customer Intimacy and Other Value Disciplines. *Harvard Business Review, January-February*, 84-93.

Trotta, G. (2003). Dancing Around EAI 'Bear Traps'. Retrieved May, 15th, 2007, from http://www.ebizq.net/topics/int_sbp/features/3463.html

Zhang, D., Chen, M., & Zhou, L. (2005). Dynamic and Personalised Web services composition in E-Business. *Information Systems Management, 22*(3), 50-66.

Section III
VLITP Implementation Problems

Chapter X
Business Process Management

ABSTRACT

The chapter seeks to advance the practice perspective of VLITP by drawing attention to individual, collective sub-teams and host organizational sense making. It reveals some of the inner workings of VLITP implementation strategies in practice today and attempts to form the theoretical bases for examining a case for BPM in VLITP implementation situations. The chapter looks at various BPM concepts including BPM practices in project management, since the mid 1990's, though originating from the early 1920s. It introduces three waves of BPM throughout the years before providing a comprehensive definition of BPM in the VLITP situation.

INTRODUCTION

Information management is a critical tool and probably the most important resource for managing a VLITP. The slogan 'information resource management' was at its peak during the early 1990's, at which time it was based on the axiom that information is an asset with a value, and needed to be managed, like any other asset, to maximise its value across the entire organization (Angell and Smithson, 1991). Currie and Guah (2006) show this principle to still be true for VLITP when managers have the capability to improve the effectiveness of certain part of the project or even gain some advantage by using additional information. On the other hand, if certain types of information about the project are not available to all project managers the competitive advantage can be transferred to those in position of such information. An example could be that a certain piece of information relating to a crucial stage of the project was only revealed to a limited number of project managers. That strategic decision to exclude the other managers gives the selected few competitive edge and can be described as information resource management.

Mismanaging project information for a VLITP could affect the achievement of its objectives that has been set. This can lead to a course of actions that misinterpret the actual plans and how the management team hope to attain the project goals. Proper planning of VLITP prevent the investment being at the mercy of management whims and external pressures which often leads to VLITPs loosing sight of their objectives. This is an aggregate of the significantly low level of forward planning by businesses against the often far too rigid planning in the public sector. The private sector businesses are content with carrying on operations virtually on a day-to-day basis. On the other hand public service organizations take very little account of likely and possible political and economic changes.

BUSINESS PROCESS MANAGEMENT

BPM involves the holistic approach of all systematic attempts to control and improve the implementation of a business process. This is why BPM encompasses optimization of individual activities, optimization of process flow, process change management, and change management for the organizational culture. The IT industry has developed BPM software that is based on the business improvement life cycle. This cycle shows the organizational desire to constantly improve the implementation of the business process. BPM software should in theory enable analysts to model, simulate, deploy, execute, measure the performance, and analyze new process implementations. However this is currently not possible, because the BPM software is not able to change the process flow, which is embedded in the applications. Another problem is that the current modeling environments are too technical for the business process analysts.

BPM is the philosophy of how business processes should be managed. This management is frequently misinterpreted as the streamline of delivering a document to the appropriate recipient (Keen, 2004). Pritchard and Armistead conclude from a series of interviews that managers see BPM as a holistic approach for process performance in the long run including process commitment (Pritchard and Armistead, 1999).

As mentioned earlier there currently is much discussion about the definition of BPM. Lee and Dale (1998) present a good overview of this subject, though it does not encompass recent changes to process thinking. Their definition for BPM is: "a customer-focused approach to the systematic management, measurement and improvement of all company processes through cross-functional teamwork and employee empowerment". Our objections to this definition are that terms like cross-functional teamwork and employee empowerment refer to an approach to BPM. We would like to have a general definition that also fits other approaches.

Our definition of BPM is based on Pritchard and Armistead (1999) findings and the research done by Smith and Fingar (2002). We have defined BPM to involve the holistic approach of all systematic attempts to control and improve the implementation of a business process. It is an holistic approach that encompasses both the soft and hard elements of the process implementation within an organization as mentioned in the introduction of this chapter. However, not all changes to the business process implementation are BPM. The systematic attempts indicate that the current implementation or changes to that implementation have been planned. The word attempt also indicates that the actions taken to control or improve a process implementation can be unsuccessful. Finally the word control is used to indicate that an organization has to monitor the current process implementation.

The holistic approach refers to the combination of the hard and soft elements in process management. The hard elements represent tangible concepts such as process modeling and process monitoring, while the soft elements represent human aspects like an employee's motivation. The misinterpreted view leaves out this human aspect, forgetting that every change in a process will affect the lives of the employees involved with it. More importantly, it does not take into account that employee commitment is essential for a project's success (Corrigan, 1996).

The reason why human aspects get a disproportionate amount of time is that they are hard to generalize. Generalization contradicts the "human touch", which is expressed differently by every manager. The lack of generalization encourages high management to delegate the soft elements to lower management, which has become responsible for passing on and executing the new organizational policies. High management often forgets employee involvement in their decision making after delegation.

BPM encompasses optimization of individual activities, optimization of the streamline between the activities, process change management, and change management within organization culture. Kretzers (2006) mentions that BPM has implications for the following four aspects:

- **Strategy:** The strategy of the organization must explicitly be reflected in its processes. They will provide a shorter link between strategy and operations, without the influence of territorial and functional managers.
- **Governance:** new compliance laws such as the 404 section of the Sarbanes-Oxley Act (SOX) require that responsibility of business processes and policies is taken by management on the highest level, including the CEO, and for sub processes at the departmental level.
- **Organization:** The organization structure needs to recognize the interdependencies and relationships between departments to stimulate value creation. The structure needs to empower employees to seek improvement across organization boundaries.
- **Culture:** The business culture needs to be ready for rapid change. A culture will be present where constant change is needed to stay in step with fluctuating business conditions. The methods, procedures and skills that support all stages of the process life cycle must reflect these culture needs.

The remainder of this chapter focuses on the hard elements. Although the human aspects should never be forgotten, they are difficult to describe without mentioning the obvious like "one should involve employees to create commitment". The next section describes the history of BPM. This is followed by Business Process Management Systems, which is software providing business process modeling, simulation, execution, and control.

THE TREND

BPM is not a new concept. The term has first been used in the mid 1990's, but originates from the early 1920s. The general concept of what it encompasses has changed throughout the years and still brings much discussion. First we will introduce three waves of BPM throughout the years as discussed by Smith and Fingar (2002) and then we will provide our definition of BPM.

First Wave

The concept of the first wave was dominated by Fredrick Taylor's theory of (scientific) management. He was able to prove that it was more efficient to standardize processes than to look for an extraordinary man. He observed how activities were performed and measured them in time and quantity. This enabled him to find the best way to perform the activity. Taylor introduced standardization around the same time as Ford introduced the assembly line (McGoveran, 2004).

Standardization had led to a drastic improvement in product quality and productivity. Taylor's scientific management also applied to management ranks. In his opinion managers were responsible for motivating and monitoring their employees. In order to do this effectively, specialization in management was required. This view had led to the introduction of managerial departments like accounting, recruitment, and production. The supervisor of a department became responsible for the available resources and had to report to a Chief Executive Officer, which in turn was responsible for the business goals and the organization's vision (Chang, 2006).

IT emerged during the first wave. It was used to automate business activities in order to reduce labor costs, improve consistency, or increase volumes. The initiative to invest in IT was made by the individual departments, which resulted in a lack of alignment with other departments. Organizations were building a fragmented infrastructure with many standalone applications and a large overhead in functionalities and data.

Second Wave

The second wave of BPM started with the publication of two academic articles in 1990 involving Business Process Reengineering (BPR). In the first article the authors argued that business processes could be improved by redesigning them and combine it with the usage of IT (Chang, 2006; Davenport and Short, 1990). They had defined BPR in five steps:

- Setting business objectives
- Identifying the processes that have to be redesigned
- Measuring the existing process
- Identifying the possible contribution of the introduction of IT to the process
- Making a prototype of the new designed process

The revolutionary aspect from their approach was that they perceived that IT should not only support a single business activity, but also the whole process. Applications had to be responsible for the process flow, meaning that applications could also initialize business activities present at other applications. In the second article, which was published around the same time as the article of Davenport and Short (1990), Hammer (1990) described his own view of BPR. Hammer (1990) claimed that the automation efforts of the past did not improve productivity significantly. The first wave of BPM the process was improved incrementally, however Hammer suggests to rebuild it from scratch (Chang, 2006). The benefits of this approach can be explained by a local optimization problem. Fine-tuning a process resembled looking for a local optimum, whilst rebuilding it enabled exploring other available local optima. The other local optima could yield higher revenues than the initial local optimum. Thus the only way to ensure the best possible implementation of a process was starting with a clean sheet.

The approach to BPR promoted by Hammer (1990) was more popular than Davenport and Smith 1990). The increased demand for system integration had led the IT industry to building huge Enterprise Resource Planning (ERP) packages. These modular build systems were implemented throughout the organization. The advantage of ERP was that ERP regulates the process flow internally. ERP vendors advertised their products as all the IT support you ever needed in a single system. This view was much exaggerated, because many business activities were still performed by specialized software. These systems had been built based on more than 20 years of experience and expertise.

Third Wave

Currently we are in the third wave of BPM, which has started in 2002 with the publication of Smith and Fingar's book Business Process Management: The Third Wave. This wave is characterized by the realization that business processes require constant improvement called agility. Organizations need to have agile processes to be able to adapt to their changing environment. During the second wave many business processes has been reengineered to make them fit an upcoming implementation of ERP. However the lack of agility of ERP prevents us from being able to have a constant reengineering approach. This lack of agility is often compared with concrete: moldable until it is solidified. The organization can mold their process and configure ERP in the implementation phase, but once it has chosen a certain implementation it is unable to change it. The reason why it is so hard to change ERP is the complexity of the related processes. Changing a simple activity at the delivery department could successively have great consequences for its ledgers, the ordering process of raw materials, and production planning.

The solution provided by the IT industry has been like it has always solved complexity problems: adding an extra abstraction layer. A Business Process Management System (BPMS) is used to separate the process flow from the execution. Existing applications perform business activities like they have always been doing during the first wave, while a new application, the BPMS' orchestration module,

Figure 10.1. Timeline showing the three BPM waves

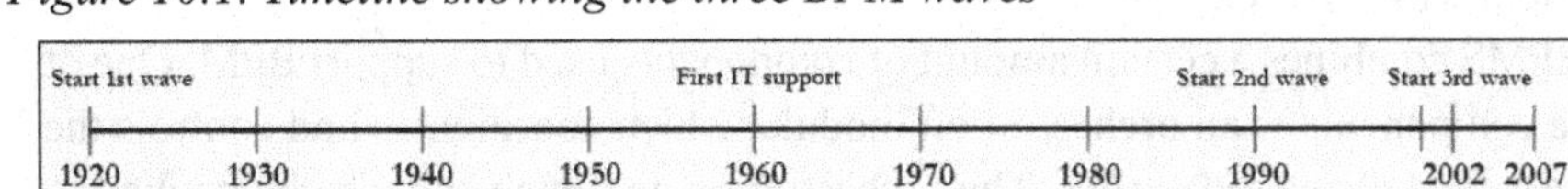

coordinates and controls the process flow based upon a set of business rules. An example of this separation is given in the next section.

The three waves of BPM are illustrated on a timeline in Figure 10.1 and can be summarized as:

- **First wave:** Standardize activities to achieve better results. IT during this period is used to optimize single business activities.
- **Second wave:** Not only improve single activities, but also the entire process flow. Re-engineer business processes to a way that fits the technology.
- **Third wave:** Agile business processes. The process flow is separated from its execution. Technology must fit the business process.

BUSINESS PROCESS MANAGEMENT SYSTEMS (BPMS)

As explained earlier, the current philosophy is that organizations have to constantly reengineer their business processes to be able to deal with the changing environment. Most of the current applications are not able to provide this agility, because they have been designed to perform a specific set of tasks. Changing the process implementation also affects the desired support from IT. The applications can't provide the desired agility, because it is difficult to change the process flow and to provide the required data.

This can be illustrated with a small example. There is a store that runs several applications responsible for specific tasks like inventory control (IC), financial information (FI), and customer relation management (CRM). A salesclerk enters a new order into his computer, which is validated by the running application. It sends a message to FI and CRM when the order has been approved, informing those applications that the customer has bought some products. FI updates the appropriate ledgers and forwards the message to IC. IC now checks the current stock levels and informs FI when it has to order products to keep stock levels within predefined levels.

The example illustrates that the decision to inform an application is made in the code of the individual applications. This is undesirable, because it would require scanning every application to find all links between the systems. Having these links is essential, because if the implementation of an application changes then the designer has to test all affected processes. It also makes it more difficult to replace a system with another.

BPMS combines a certain amount of components used to support BPM. One of these components is an orchestration module, which coordinates and controls the process flow discussed earlier. The orchestration approach can also be applied to the previous store example. The sales clerk still registers the order using his computer. This computer sends a message to the orchestration module, which initializes the sale process. First the orchestration module will request to validate the order. When approved it sends a message to CRM, FI, and IC. Thus the BPMS module is responsible for the process flow, while the other applications provide specific functionalities. More precise: the other applications provide services!

A BPMS offers more than just an orchestration module. It is involved with every phase of a process' improvement life cycle. The nine phases of business process improvement life cycle are shown in Figure 10.2. The current available BPMS applications require involvement of a business analyst and a technical analyst. The business analyst specializes in how process should work, but often lacks technical understanding. This is where the technical analyst comes. This technical analyst can translate the process model of the business analyst into a model which is used by the BPMS.

When closely examined, one sees that the improvement life cycle is actually an elaborated version of the well-known Deming cycle (also known as Plan-Do-Check-Act, PDCA, or Shewhart cycle). The improvement life cycle is initialized with discovering a business process. During this phase the organization becomes aware of the individual activities within the process and their relationship. It then registers the event -, message -, and control flow (Smith et al, 2002). Discovering helps to get a clear picture of how a process is really implemented, instead of looking at a usually outdated formal description. Furthermore, it usually reveals possible changes, because employees of all levels of the organization are consulted.

The next step in the life cycle is the design phase consisting of defining, modeling, and simulating a redesigned process (Figure 10.2). Defining a process is creating a description with clear inputs and the desired outputs. The description itself is less useful than the process of making it, because writing a formal description helps to create a common view of the process among the involved employees.

Figure 10.2. Process improvement life cycle (Snur, 2006)

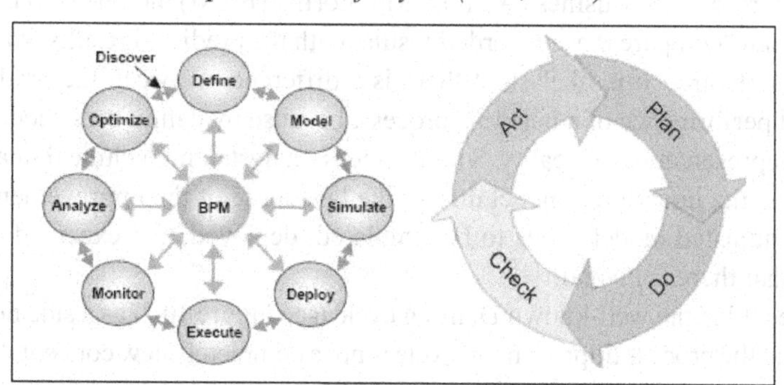

Conceptually rebuilding the process is done at the modeling phase. Many BPMS tools provide a graphical modeling environment, which allow analysts to make a visual representation of the process like they would on a sheet of paper. The resulting model consists of business activities, their relationship (the process flow), and performance indicators. Embedded simulation software can be run after finishing a model. The implemented performance indicators now provide data about the performance of the corresponding activities in terms of queue lengths in time and quantity, the number of products that have been processed, or failure rates. The analyst tries to optimize the performance indicators by altering his model allowing him to observe different scenarios without really having to implement them.

The next phase in the improvement life cycle is deploying the designed process model. In theory this should be quite easy to implement. BPMS software has to be able to export a process model to a set of executable business rules and the business activities should be implemented as services enabling them to be rearranged without additional programming. However this ideal situation does not and probably will not ever exist. Current BPMS software is not able to make the translation between the process model and the business rules without the input of a technical analyst. Furthermore, most applications are not Service Oriented and those that are still require some adjustments.

The model is usually first deployed in a test environment and employees are trained for their new responsibilities. These preparations do not have to affect the current process and minimize the risk of its continuation. Management has to ensure that every involved participant including man and machine is ready to "go live" with the execution phase. The newly designed process is implemented throughout the organization and replaces the old version. The BPMS orchestration module uses the new business rules to coordinate the various activities.

The execution phase does not end when the next phases start, because the process execution will only stop when a newer version of the business process is implemented. Executing, monitoring, and analyzing are therefore performed parallel with each other. Process monitoring consists of recording real performance of each activity by BPMS' Business Activity Monitoring (BAM) module. A business analyst can now compare these recorded results with the predicted results from the simulation software. It is likely that there is a difference between the predicted and the real performance of a business process, because by definition a model is a simplified representation of reality. The business analysts and technical analysts will optimize the underlying model during the last phase of the improvement life cycle. This adjusted model needs to be simulated, deployed, and executed again indicating that there is no end.

We have added the well-known Deming cycle (see Figure 10.2) as a side note to illustrate that the process improvement cycle is not a completely new concept. It was first introduced by Shewhart in 1939, but only became famous with the publication of Deming's book Out of the crisis in 1986. It consists of a four-step approach to model continuous improvement in the design of products, processes, and services. They directly correspond to the phases in Figure and consist of plan (define, model, simulate), do (deploy, execute), check (monitor, analyze), and act (optimize).

BPMS Characteristics

Until now we have only discussed what support BPMS gives to the process improvement life cycle. Hill et al (2006) suggests a BPMS should encompass the following:

- Orchestration engines coordinate the sequencing of the activities and steps (system and manual) according to the flows and rules in the process model;
- Business intelligence and analysis tools support analysis of data produced during process execution. Capabilities range from reporting to online analytical processing analysis to graphical user dashboards. Business activity monitoring (BAM) systems do this in real time with proactive alerting;
- Rule engines execute rules that abstract business policies and decision tables from the underlying applications, and make available more-flexible process changes;
- Repositories contain process definitions, process components, process models, business rules and other process data to enable reuse across multiple processes;
- Simulation and optimization tools enable business managers to compare new process designs with current operational performance. Scenarios are executed, altering resource constraints and business goals, to assess risk and display the financial and operational (that is, timeliness and quality) impact on the organization;
- Integration tools link the model to other system assets (data and logic) that support work steps.

This however, does not include components, but represents the full overview of the capabilities of a BPMS.

BPMS Standards

A well-known joke among developers is that the beauty of standards is that there are so many to choose from. Choosing a non-surviving standard is very expensive, because it will require converting all of the software that has been based on the non-surviving standard. The lack of real standardization has made organizations reluctant to invest in the past. When looking at BPM standards one might conclude that IT industry has learned from their mistakes eventually.

The IT industry had made a bad start by introducing many process modeling languages like: XLANG, WSFL, WSCI, BPML, BPEL4WS, BPSS, and XPDL. The two most popular languages were the Business Process Modeling Language (BPML) and the Business Process Execution Language for Web Services (BPEL4WS, or BPEL in short). BPML was developed by the Business Process Management Initiative (BPMI) and was originally supported by major vendors like Sun, BEA, Intalio, and SAP. However, four years later the BPMI favored the use of BPEL, after merging with OMG (Ghalimi, 2006). Therefore BPEL is considered the dominant process modeling language standard.

BPEL is an XML based language, which orchestrates the message flow between the business activities. Actually it combines the directed graph based process of IBM's WSFL and the block structured process of Microsoft's XLANG (Moon et al, 2004). Officially BPEL can be used to model two types of processes: executable and abstract processes. An abstract process is a business protocol, specifying the message exchange behaviour between different parties without revealing the internal behaviour for any one of them. An abstract process views the outside world from the perspective of a single organization or service. The message exchange for abstract processes is called choreography, which tracks the message sequences between involved parties. An executable process views the world in a similar manner as abstract processes, however, things are specified in more detail such that the process becomes executable, for example an executable BPEL process specifies the execution order of a number of activities constituting the process, the partners involved in the process, the messages exchanged between these partners, and the fault and exception handling required in cases of errors and exceptions. The message exchange for executable processes is called orchestration (Peltz, 2003).

Figure 10.3 demonstrates the differences between executable and abstract processes using orchestration against choreography (Peltz, 2003). Several ques-

Figure 10.3. Orchestration versus choreography

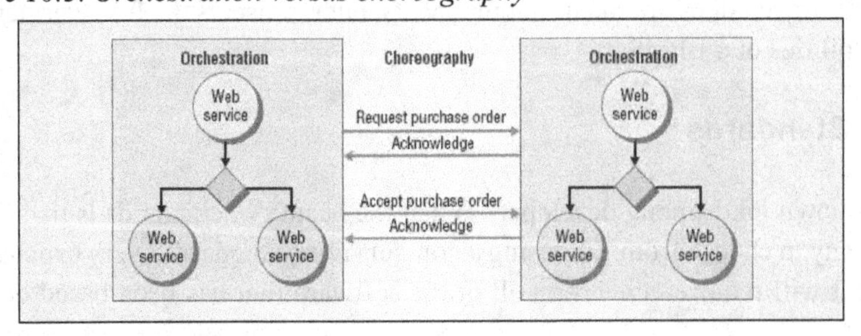

tions have been raised concerning the suitability of BPEL for modeling abstract processes or even if BPEL should be considered a programming language rather then a modeling language. Although the goal of SOA and BPMS was giving the process back to business, this should not lead to thinking in terms of programming only. Vendors have been developing tools to graphically represent BPEL code, but these tools only reflect the XML basis. Consequently, the project members need to think in BPEL constructions.

A more suitable modeling language for VLITP is the Business Process Modeling Notation (BPMN)—a graphical notation standard for representing business processes that developed by the Business Process Management Initiative (http://www.bpmi.org). It is able to graphical represent a business process without having the constraints found in the graphical BPEL variants. There has been some research in transforming BPMN models into BPEL (White, 2005), but unfortunately we have not been able to find a tool that is able to fully transform a BPMN model to BPEL. There are many tools available that give a basic representation, which then has to be modified and optimized by a more technical member of the project team. This explains the need for a business process improvement life cycle to be both business and technically oriented analysts (see previous sections). While the gap between business and IT may seem to be narrowing, there still exists gap.

VLITP MANAGEMENT PROCESS STAGES

This section describes and organizes the work of a VLITP into six process stages which are interrelated, depends on each other and iteratively (Figure 10.4):

Project Initiation

This is the beginning process for all VLITPs where decision about undertaking the project is made by examining the costs and benefits of the particular project to the host organization. This stage may also include the comparison of one project to another. A positive final decision at the end of this stage would lead to the commitment of resources for the particular project. This stage is therefore the formal recognition that a VLITP should begin. Some of the resulting documents usually produced at the during the initiation process are:

- Definition of particular goals and objectives for a VLITP.
- Evaluation and determination of VLITP benefits.
- Selection of a VLITP based on criteria defined by a selection committee.

Figure 10.4. Stages for VLITPs

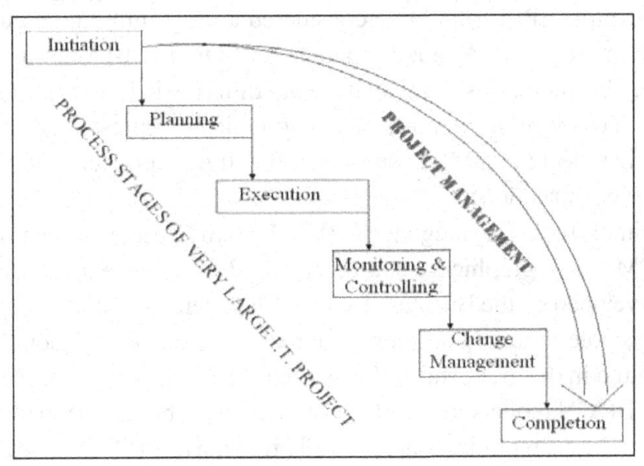

- A set of written charters for a particular VLITP.
- The assignment of a project manager.

Project Planning

The planning process stage is the heart of all successful VLITPs. Proper planning techniques can be the difference between a failed project and a successful one (Willcocks and Lacity, 2006). This stage involves an outline of the essentials products for completing the work for VLITPs. This function justifies the importance of this stage of a VLITP management process and can consume a large amount of the overall project time, but it's well worth the investment.

This stage may involve researching, communicating and documenting which determines how the project will progress through the remaining processes and establishes the foundation for the rest of the project. It is very important to communicate well with the stakeholders through this process, assure that all project team members and stakeholders understand the purpose of the project and how the work will be carried out, and establish a professional decorum with everyone involved, the stakeholders will feel confident that the project will be successful (Angell and Smithson, 1991). These activities will gain their cooperation later in the project when the problems start to appear.

The Planning stage (Figure 10.4) is the largest of all other processes and usually produces the road map for achieving the goals a particular VLITP was undertaken to address. Some of the resulting documents usually produced at the during the Initiation Process stage may include:

- Determination of VLITP deliverables and milestones.
- A written scope of statements for VLITP.
- A list of requirements for a VLITP.
- A break down of the work of the VLITP into tasks.
- The development of a project schedule.
- The establishment of a budget for the VLITP.
- The development of a risk, communication, quality and change management plans for the VLITP.
- The determination of a resource needs for VLITP.
- The assessment of special skills needed for the VLITP tasks and identification of resources.
- A document setting the stage for the success of the VLITPs.

Project Execution

This is where the real work of the VLITP actually happens. After great plans have been put together to determine the project success, Execution process stage (Figure 10.4) is required to it follow-through and put such good plans into action enabling the project team members to complete the tasks. This stage keeps the VLITP team focused on the work and communicates project progress to stakeholders and management. After the work of the VLITP begins, sometimes the project plan needs to be change. At this stage of project management process the project managers can update the project planning documents and redirect and refocus the project team on the correct tasks. The execution process is the stage where majority of VLITP resources, spend most of the project budget and run into scheduling conflicts.

Some of the resulting documents usually produced at during the Execution Process stage may include:

- Obtaining VLITPs resources.
- Established VLITPs team.
- Direction and lead to VLITPs management team.
- Written status reports and other information or VLITPs.
- Communication of VLITPs information.
- Management and direction contractors for VLITPs.
- Management of VLITPs progress.
- Implementation of quality assurance procedures for VLITPs.

Project Monitoring and Controlling

This stage of project management processes involves monitoring the work of the project and taking performance measures to assure that the work performed is on track with the project scope and that the deliverables are being met. If performance checks during this process show that the project has veered off course, corrective action is required to realign the work of the project with the project goals. Corrections and changes during this stage of the project management process may require a trip back through the Planning and Executing process stages. Most often this will occur either for the purpose of change requests or for corrective actions.

Some of the resulting documents usually produced at the during the Monitoring and Controlling stage may include:

- Measuring performance and comparing to VLITP plan.
- Ensuring that the VLITP progresses according to plan.
- Taking corrective action when measures are outside limits.
- Evaluating the effectiveness of corrective actions.
- Reviewing and implementing change requests.
- Updating the project plan to conform to change requests.

Most VLITPs tend to skip the Closing Process Stage simple because the project management team find it easy to focus on the next project one the project at hand is completed. However, there are important functions to be performed at the end of the project such as: obtaining sign-off of the VLITPs, document lessons learned and close out a project that's complete and that stakeholders really love the final products.

The documentation of lessons learned is the most important aspects of stage of the project management process. That's because at this stage the project management team would realise that a VLITP has just been completed successfully where some processes have worked very well and others could have been improved. Now is therefore the time to capture the good and the bad so that the next VLITP to be undertaken can capitalize on the lessons learned during this project. Though the most popular aspect of this process is celebration, usually because the project management team has met or exceeded the agreed-upon project goals and the stakeholders are satisfied. Celebration usually means success and success are nearly always celebrated, because VLITPs are truly team efforts and always appropriate to congratulate your team on a job well done.

Some of the resulting documents usually produced at during the Closing stage may include:

- Obtaining acceptance of VLITP deliverables.
- Securing sign-off from all stakeholders of VLITP.
- Documenting lessons learned from a VLITP.
- Archiving VLITP records.
- Formalizing the closure of VLITP.
- Releasing VLITP resources.

Change Management

VLITPs are usually considered an important part of a revolution taking place that could transform the organization and the delivery of its products and services. Whatever the merit of the origins of a VLITP, the fact is that it will result in the invention of new ways and methods of doing things in that organization. The expectation is that the project will affect some or all of the following changes:

- **Expectations:** This is where a VLITP is expressed in way that it takes the organization from a state of stability to a state of flux. Such expectation may lead to staff thinking they could be working for another organization or in a completely new occupation or profession. Some staff get to believe a VLITP will bring them an ever-grater diversity of choice and ever-greater increase in the quality of what is actually on offer.
- **Social:** Both staff and customer of the organization implementing a VLITP may consider it to change their lives in many ways. These may include lifestyle choice, the ability to buy and possess certain items, the ability to benefit from an increase in quality of certain items or service, or even deliver a long awaited item or service.

The management team of a VLITP is always aware that every change that occurs to the project stages affects the total outcome of the project. The effect could be either positive or negative. The team must keep alert to the fact that most changes to the project add a lot of expense and consequently an increased risk to the project. Risk could either be that the team lose track or the scope may creep up sooner than expected. Most times when changes occur to VLITPs, they result to expansion of the project. A well-organized team avoids such problem by implementing one of various forms of change management systems. An even better strategy is to follow-up with some form of education to customers about the costs of such change.

Completion

Closing a VLITP is often an ambivalent activity. Closure may be scheduled at any point after the completion of the last sub-project within the project portfolio. To a large extent, the exact time a VLITP formally closes will depends on the amount of support required to ensure the new operational environment delivered by the systems is fully embedded. This requires the systems to achieve outcomes and realize benefits.

However, VLITP closure identifies the need for future assessment of benefit realization as well as a formal review of those achieved so far. The project team carries out a formal confirmation process to ensure the Business Case for the particular VLITP has been satisfied. Occasionally such investigation proves new system benefits to be like profits—one simply wouldn't know their real value until the last dollar has been accounted for. This makes the closure open to a lot of speculation, with the potential for undue optimism. We are inclined to think that perhaps a VLITP should remain open for a specified period after the project trenches have been completed. This is to ensure that there is impetus to garner all of the planned benefits and that the program is fully accounted for.

An interesting area of VLITP management that has not been well documented and probably the most crucial step in achieving a successful outcome is, the 'launch of products of a VLITP or, more specifically, the "transfer of the care, custody and control" of products of a VLITP into its working environment. Popular literature contains a rather vaguely reference to the need for training and supporting products of VLITP, but this tends to be somewhat brushed aside as a separate activity. What this book emphasizes as an important part of a VLITP completion stage is the need for "marketing and selling" of products that emerged from a VLITP into its working environment. While this activity might be well established in the commercial or retail sectors, they are not always essential activities if the full benefits of a VLITP emerging products are to be garnered within the framework of managing the project portfolio. Considering the acceptance of project delivery by the stakeholders is the first step to "stakeholder satisfaction", the closure of VLITP is a very vital part of VLITP management strategy.

VLITP management is about the successful delivery of acceptable "deliverables" in terms of achievement within constraints of time and resources. Efficient VLITP management goes much further and requires optimum selection of all sub-projects in the first place and the realization of intended benefits in the final stage.

CHAPTER SUMMARY

Properly managing VLITP information can help ensure compliance with regulatory requirements for retention of critical information (Reason, 1997), provide properly documented audit trails, maintain clear and permanent records of both current and historical information, as well as providing an efficient means for quickly identifying, searching, finding and retrieving information relating to various project objectives.

Business Process Management (BPM) involves the holistic approach of all systematic attempts to control and improve the implementation of a business process. Our understanding of BPM has changed throughout the years. There are three waves of BPM each referring to a specific interpretation of how BPM can be applied. These waves have the following characteristics:

- **First wave:** Standardize activities to achieve better results. IT during this period is used to optimize single business activities.
- **Second wave:** Not only improve single activities, but also the entire process flow. Re-engineer business processes to a way that fits the technology.
- **Third wave:** Agile business processes. The process flow is separated from its execution. Technology must fit the business process.

It is our current understanding that process implementations have to be constantly improved. Working with the business process improvement life cycle ensures that these constant improvements are being implemented. The IT industry has developed software to help organizations model and simulate new implementations called Business Process Management Systems (BPMS). These systems are also able to deploy, execute, and monitor the new process implementation depending on the existing IT architecture.

Communication has been shown throughout this book as a vital tool for the management team of VLITPs. That is because without good communication all attempts to carry on the various parts of the project will definite fail.

REFERENCES

Angell, I.O. and Smithson, S. (1991) *Information Systems Management: Opportunities and Risks,* Macmillan Press, Basingstoke.

Archibald, R. D. (2003). Managing High-Technology Programs and Projects, third Edition, Wiley, p.44.

Brown, A.D. and Jones, M.R. (1998). Doomed to Failure: Narratives of Inevitability and Conspiracy in a Failed IS Project, Organization Studies, Vol.19, No.1, pp.73-88.

Brown, T. (2001). Modernization or failure? IT development projects in the UK public sector. *Financial Accountability and Management, 17*(4), 363-381.

Currie, W. L. & Guah, M. W. (2006). Web Services in National Healthcare: The Impact of Public and Private Collaboration. *Information Systems Journal, 1*(2), 48-61.

Eyre, E. C., & Pettinger, R. (1999). *Mastering Basic Management.* Third Edition, MacMillan Press, London.

Galliers, R. D. (1998). Reflections on BPR, IT and Organisational Change. In *Information Technology and Organisational Transformation: Innovation for the 21st Century Organisation,* (Galliers, R. D. and Baets, W. R. J. eds) Wiley, Chichester, pp. 225-243.

Guah, M. W., & Currie, W. L. (2003). ASP: A Technology and Working Tool for intelligent Enterprises of the 21st Century. In *Intelligent Enterprises of the 21st Century.* Edited by Gupta, JND and Sharma, SK. Summer. IGI Public, pp.188-219.

Kerzner, H. (1989). Project management: A Systems Approach to Planning, Scheduling, and Controlling, 3rd Edition, Van Nostrand Reinhold, NY, p84.

Markus, M. L. (1994). Electronic mail as the medium of managerial choice. *Organization Science, 5*, 502-527.

(Moon, Lee, Park, and Cho, 2004)

Patel, M. B., & Morris, P. G. W. (1999). *Guide to the Project Management Body of Knowledge.* Center for Research in the Management of Projects, University of Manchester, UK, p. 52.

Reason J. (1997). *Managing the risks of organisational accidents.* Aldershot: Ashgate.

Reddy, R. (2004). Cultivating the IT Portfolio Manager. *Cutter Consortium Business-IT Strategies Executive Report, 7*(1), January.

Smith, D. (2000). E-business strategy risk management. *Computer Law and Security Report, 16*(6), 394-396.

Stewart, R. (1997). The Reality of Management. Third Edition, Butterworth-Heinemann, Oxford.

Youker, R. (1989). Managing the project cycle for time, cost and quality: lessons from World Bank experience. *Keynote paper, INTERNET 88, Glasgow, 1988, 7*(1) February, p. 54.31.

Chapter XI
Outsourcing and Escalation Issues in VLITPs

ABSTRACT

VLITP escalation has been documented to be a widespread phenomenon in the 21st century. Nearly every research in this area has portrayed escalation as an irrational decision-making process whereby additional resources are plowed into a failing project. This chapter examines the possibility that some of these escalation issues could be appropriately managed by avoiding irrational actions and rationally responding to various situations that may occur in a VLITP. Later on in the chapter, the author disperses popular belief that VLITP outsourcing is in the nature of partnership and strategic alliances. It exposes that VLITP outsourcing vendors do not share the same profit motives as the host organization who is meant to enjoy the benefits of the VLITP and therefore baring the full cost of the implementation. It further suggests that a tight contract is the only mechanism to ensure that expectations of the host organization are met. Host organizations must negotiate and agree that the contract contains a number key issues including a successful outsourcing relationship between the host organization and the outsourcing vendor.

INTRODUCTION

Outsourcing some or all of VLITP is a common policy being encouraged by host organizations mainly because it saves money and human resources. But if outsourcing is not managed properly, it also could lead to expensive mistakes, unmet expectations and even failure of the VLITP (Lacity and Willcocks, 2006). There are a several issues involved in outsourcing VLITP that can go dramatically wrong. VLITP outsourcing is the subcontracting of a part or all of the VLITP implementation by the host organization to one or more an external IT service provider (Guah, 2008; Mahnke et al, 2008). The degree of subcontracting involved may varies across the whole spectrum ranging from just one part of the programming task to the wholesale outsourcing of an entire business function of the host organization. Due to the need to reduce cost and maintain continuous performance by the host organization while implementing a VLITP, outsourcing has gained tremendous momentum in the VLITP environment.

The common practice by most host organizations in regards to outsourcing contracts is the passing of one or more implementation, infrastructure management and/or operational responsibilities from an in-house IT department to an outsourcing vendor through a contract. Such contract quite often involves complicated business and legal issues, and is fraught with risks for both the host organization and the outsourcing vendor. Lacity and Willcocks (2006) suggest that outsourcing vendors and the host organization are not partners because their profit motives are not shared. As a result the host organization cannot expect the outsourcing vendor to act in the best interest of the host organization in situations where a conflict of interest arises. Such situation necessitate a written outsourcing contract; thus, making the contract the most important instrument for defining the fights, liabilities and expectations of both parties which guides the behaviours of both parties concerned (Agrawal et al, 2006; Lee, 1996). It is therefore important for top management of host organization to have some understanding of the complicated business and legal issues involved in VLITP outsourcing and have some awareness of how these issues should be addressed in the contracts concerned.

An outsourcing contract often includes a collection of related agreements covering a variety of issues such as service level, transfer of assets staffing, pricing and payment, warranty and liability, dispute resolution mechanism, termination, intellectual property matters, and information security—very extensively explored by Lee (1996). More than a decade later, very few outsourcing contracts in VLITPs are problem free. There seems to be simply too many things that can go wrong, too many parts of the VLITP that need to be carefully overseen, and too many aspects of the contract to be delicately negotiated to prevent the host organization from becoming mired in some sort of management issue (Di Romualdo and Gurbaxani, 1998).

POSSIBLE PROBLEMS WITH VLITP OUTSOURCING

When a host organization is reviewing a potential outsourcing contract involving a VLITP, few issues of potential problems should be investigated. The first issue of contention is usually making sure the organization (and not the outsourcing vendor) will end up owning the source code of any IT application resulting from the VLITP (Johnson, 1995). The contract must be flexible enough that the service being outsourced can change as the business changes (Mahnke et al, 2008). The project champion should be able to answer these simple questions:

- Is there a system set up to gauge the success of this VLITP?
- Is there a clearly defined—preferably one—person established as the liaison and can that person speak the language where the work is being done?

An increasing number of government departments and private multi-nationals all around the world are using outsourcing for VLITP implementation today, mainly to reduce or stabilize costs, access advanced technology, compensate for a lack of skilled IT workers, improve business efficiency, and remain competitive in the global marketplace. Unfortunately many executives assume the outsourcing vendor shall come in and takes over, thereby allowing in-house staff to wash their hands off the VLITP (Ranganathan et al, 2007). The actual process of implementing VLITPs almost never works out that easily, probably explaining why many statistics on outsourcing success and failure claim about half of all outsourcing engagements in IT projects fail. One undisputed fact is that the relationship between host organizations and outsourcing vendors requires quite a bit of communication and a great deal of flexibility.

Many host organizations run into trouble when they choose a vendor based solely on pricing variables alone (Dos Santos, 1991). It follows the common saying "you get what you pay for". Understandably, companies are much more conscious of cost cutting in the current economy. However, making such an important decision about an organization's future purely based on price is a very big mistake. Closely related to total price of implementing a VLITP is the responsiveness of the vendor's to the continuity of the overall goals. Responsiveness of the vendor's customer service is incredibly important and should the premise for a guaranteed in service-level agreements. If the outsourcing vendor has expertise in the industry of a specific VLITP (i.e. healthcare, military, airline, etc.) the contract will demonstration a higher level of flexibility (Agrawal et al, 2006) That's because the outsourcing vendor would want to reassure the host organization that the quality of the final outcome will be in no doubt satisfactory. That involves make sure that relevant staff of the host organization understand how to use the emerging technologies to move their functions in the business to the next level and attain better results which ultimately leads to dramatic improvements to future business goals."

Considering such magnitude to effort required for a VLITP outsourcing contract to be right, executives have to consider if this is something can be controlled from a distance (Gonzalez et al, 2006; Ranganathan et al, 2007). Of course while several people within the host organization may be involved with the outsourcing negotiation of a VLITP, there needs to be one person in ultimate control. This should avid the shock that comes to light only after calculating the cost of the time various staff of the host organization spend managing the relationship, and realizing that it costs more than initially thought the organization could save by implementing the VLITP. One significant finding in all three cases found in this book was that in-house IT managers underestimated how much time would be required to manage the outsourcing VLITP along with outsourcing vendor relationship (McFarlan and Nolan, 1995; Shi et al, 2005). As a result, all VLITP management teams were found to be wrestling with issues relating to better managing outsourcing contracts.

Another major issue related to outsourcing VLITP is that of divided loyalties (Earl, 1996). A small firm (in Korea handling software development for a multinational with head office in Paris for example) may be most concerned with enjoying business benefiting in a local context. Executives must realise that before the VLITP was outsourced, people working on smaller IT projects were its own employees and generally trusted as such. However, after engaging in an outsourcing arrangement, there has now been a situation where people have to check the checkers because these people don't have loyalty to your company long-term interests. In-house IT managers' functional roles suddenly include spending more time on the project that was probably not anticipated.

A sample list of issues the VLITP to be considered when setting up an outsourcing relationship contains:

- Ensure the contract indicates that the host organization owns all source codes for the final application.
- Flexibility to renegotiate pricing and terms should be built into the contract.
- After settling on a strong contract, ensure there is a template that does not contribute to continuously recreating the wheel.
- The person with responsibilities for managing the relationship between outsourcing vendor and the host organization must be decided at the initial stage of the VLITP.
- To emphasise the need to properly set up this relationship, sufficient time must be given to both parties to arranged liaison by not rushing the initial stages of the VLITP (McFarlan and Nolan, 1995).

- There could be a change in IT management with the host organization as a result of the VLITP outsourcing contract. The particular person who formally ran the IT operations in-house may not be the most suitable staff member to manage VLITP outsourcing contract and related functions.
- By deciding, at the early stages, how the host organization would measure the anticipated VLITP success, there are no space for surprises (Tyler and Steensma, 1995).
- It is important to set some time aside brainstorming and discussing, with staff, what it may take to manage a third-party relationship with a VLITP outsourcing situation.
- The contract should also include strategies and/or methodologies for mediating and resolving dispute during the VLITP implementation (Earl, 1996).
- Ensure there is a clear understanding between outsourcing vendor and the host organization of specifics within the scope of the VLITP contract. This should include statements about software upgrades and (Johnson, 1995) maintenance—whether these would be paid for in addition or whether they are a part of the current costs of the VLITP.
- It is important to get a clear understanding about the outsourcing vendor's consideration of this business relationship, indicating whether it is strategic or operational. If the host organization does but the outsourcing vendor doesn't, then both parties are out of sync from the start and that call for future concerns about the final outcomes of the VLITP.

It is very important to be absolutely clear about the capabilities and expertise of the outsourcing vendor. This is often quite different from the many claims made in a good sales pitch. Remember the success or failure in achieving the final outcomes depends on specifics of the outsourcing vendor's capabilities and expertise.

LEARNING FROM ADAPTORS OF VLITP OUTSOURCING

The driving force behind the decision to outsource VLITP, in part or full, is typically cost reduction and the host organizational restructuring (Guah, 2008; Tyler and Steensma, 1995). In the implementation of VLITP outsourcing activities can range from buying a software package to outsourcing all software development, support and data center operations.

A key component to a successful VLITP outsourcing relationship is the ability to communicate and document performance results using meaningful business oriented metrics. Unlike a standard IT projects, metrics for VLITP outsourcing contracts have been difficult to document and quantify. Large outsourcing vendors for some prominent VLITP, with reasonably good records in pursuing and offering outsourcing arrangements, continue to take aggressive steps to remove certain level of ambiguity from VLITP outsourcing contracts (Agrawal et al, 2006). These organizations are utilizing software metrics based on function points, defects, problems and effort to document performance. The author has listed below a number of basis issues to consider during VLITP contract negotiations, targets monitoring evaluation of final outcomes.

Taking the Decision

The decision by a host organization to partially or fully outsource VLITP is not only difficult but also quite often based on frustration rather than facts. This book suggest that such an important decision that quite often determines the future of the host organization should be based on a solid understanding of capacity, expertise and performance of the IT department with the host organization (Wallace et al, 2004). Some for strategic metrics should be utilized to establish a baseline of productivity, quality and the costs associated with providing these services in-house before the final decision is made. The process of investigating such baseline metric may uncover process improvement opportunities to eliminate the on-going frustrations with the in-house IT capacity (Ardagna and Francalanci, 2005; Brockner, 1992). On the other hand, the analysis may result in the realization that outsourcing is a viable alternative for the VLITP. Having completed such baseline analysis may also allow a the host organization to better negotiate the VLITP outsourcing contract being fully aware that the desire objectives, the anticipated benefits and the metrics required to measure and manage the VLITP contract efficiently.

Establishing the Baseline

For a totally outsourced VLITP contract a broad reaching quality and productivity baseline would be much more preferred. That's because it will ensure the VLITP contract basis is a fair representation of the host organization's previous performance. Developing the baseline, on the other hand, requires a structured approach. The first thing to do in this approach is to determine baseline composition. The baseline composition should be a representative sample of the host organization's business environment. The sample should consider baseline time frames, project/application size, technology and project type—indicating whether it would be an enhancement to existing technology or a completely new development of the IT infrastructure and capacity within the host organization.

After determining the above, the host organization would need to have the baseline data appropriately collected to include: function points; financial and level of effort data; project attributes; problems and defects; and user satisfaction. The analysis of this data will provide the following valuable information to both the outsourcing organization and the outsourcing vendor: productivity rates; quality statistics; maintenance requirements; and improvement opportunities.

Such baseline results may vary significantly across different organizations, mainly due to technology, culture, history and skill base, which dictate performance and drive the baseline. Case II attempted to use "standard" quality and productivity rates to establish VLITP outsourcing contracts. It did not fully deliver the anticipated results because the use of standard rates should be approached with great caution since they have proven to be inappropriate on a number of VLITP outsourcing arrangements.

Relationship Management between Host Organization and Outsourcing Vendor

A number of issues are necessary for discussion in regards to managing the relationship between the host organization and the outsourcing vendor (Earl, 1996). Form a technology perspective it is very important for the vendor's system development methodology to meet the host organization's desire for software quality and procedure (Ardagna and Francalanci, 2005). It is imperative to understand this process because the final decision would need to ensure that the method used would fit the current development environment and allow for future maintenance (Chau, 1997). Where the VLITP outsourcing would involve the complete development function the vendor usually prefers to use its own methods and techniques (Wallace et al, 2004). The reasoning is simply because the outsourcing vendor already has the appropriate the expertise within its employment. In the event that the host organization manages to convince the outsourcing vendor to use an in-house methodology, it must be recognized that the use of unfamiliar techniques may have a negative impact on overall quality and productivity level of the final outcome of the VLITP, and may subsequently affect the ongoing relationship (Brockner, 1992; Willcocks et al, 2004). As seen in cases I and III, the typical resolution is modifications to the selected methodology are to satisfy the needs of both parties.

The next issue to be considered in at this level is the Project Management functions within the VLITP. Majority of the work associated with VLITP outsourcing arrangements are strictly project based. Due to the unique requires of a VLITP, standard project management techniques often require modifications due to the addition of the outsourcing vendor. Changes may include: tighter control over project deliverables; formal budget/schedule variance analysis; well documented roles and responsibilities for both the host organization and the vendor; and a change control process that quantifies impact on time and cost.

Probably the most debatable aspect of the VLITP contract arrangement, and requiring attention, is the accounting/budgeting process (Willcocks et al, 2004). Most software packages that deal with accounting processes in use as the 21st century unfolds were not designed with VLITP outsourcing in mind. As a result the host organization has to ensure its entire computerised accounting process be revamp for VLITP outsourcing. The accuracy of the time accounting often requires improvement as well as the incorporation of quality and productivity measures. A number of host organizations have started to push this concept further by redesigning their processes to use function points as the basis for budgeting software services by business area.

Similarly but not quite as important as the above, is to ensure there are procedures in place for adequate measurement of new applications, to effectively manage the host organization and outsourcing vendor relationship which ultimately measure the success the VLITP. Measurement does not only help to identify the need for VLITP outsourcing and is often the basis for contract payment, it also provides the primary tool for evaluating the quality of deliverables, quantifying changes in scope and identifying process improvements to the VLITP. When used appropriately, measurement could become the most effective communications tool available to both the host organization and the outsourcing vendors.

It is important to recognize that establishing the processes discussed above takes time. Many host organizations require a full year to implement these ongoing processes. Time is required to develop the processes, train personnel and integrate the techniques into the daily software development and support activities. In addition, VLITP management will need to monitor, modify and mandate as appropriate.

Validating the Results

The use of software metrics in VLITP outsourcing contracts will eliminate much of the up-front ambiguity. However, there is still the need to validate the performance of the outsourcing vendor against such agreement. Critical to this process is the ongoing validation and auditing of performance data.

The performance validation should be approached with the same level of control and discipline any business would follow to close its books, audit its results and produce its annual report. Using a cohesive approach, actual performance and variances should be documented along with specific recommendations designed to improve adherence to contract requirements (Chau, 1997). A number of host organizations have found this to be best accomplished by a mutually agreed upon, unbiased, third party. The third party would provide many services including: performance validation, function point rule interpretation, conflicts resolution and performance comparisons to industry benchmarks. This level of commitment helps ensure VLITP contract compliance and promotes the use and acceptance of software metrics as a critical component of outsourcing.

ESCALATIONS

Much of today's human factors research and expertise is channelled towards improving the ways we use information. Virtually everyone has experienced the frustration of using computer software that doesn't work the way they expect it to. For the majority of end users of information systems, if the system is not working they have no recourse but to call for technical help, or find creative ways around system limitations. This is usually in the form of using those parts that are usable, and circumventing the rest or increasing stress levels by using a substandard system. The management team of VLITPs tries to avoid this with a more complete understanding of the users' tasks and requirements had been present from the start. The development of easily usable human-computer interfaces is a major concern for agronomists within very larger projects. This also involves the design of appropriate signs, symbols and instructions so that the end users can quickly and safely understand their meaning.

An escalation takes place within a VLITP when decision makers at the host organization become over-committed to previous decisions, and invest more resources in a failing project (Zardkoohi, 2004). Such concept of escalating commitment refers to the human tendency to adhere to a course of action even in the face of negative information concerning the viability of that course of action. As seen in the Case I, decision makers at the NHSIA had continued commitment to a specific course of action despite information—by means of research evidence—suggesting that the course of action was failing (Feeny and Willcocks, 1989). Instances of escalation quite frequently involve equivocal information about project status and future prospects and that in such cases the common understanding of escalation as an irrational process of throwing good money after bad would not necessarily hold (Bowen, 1987). We also notice—from Cases I and III—the introduction of adjustments that take the form of deferring the project, switching the project to serve a different purpose, changing the scale of the project, implementing it in stages, abandoning the project, or using the project as a platform for future growth opportunities.

These actions suggest the application of expectancy theory, which proposes that individuals have a tendency to believe that allocating additional resources will eventually lead to goal attainment (see Figure 11.1). Self-justification theory explains the VLITP management's desire to demonstrate rationality to by ignoring evidence that demonstrates previous decisions might have been unsatisfactory. We see a number of these common factors in all cases of VLITP escalation situations:

- Undergoing certain amount of loss or cost—which may not necessarily be monetary—that has resulted from an original decision or course of action by the VLITP management team.
- A persisting predicament involving some continuity over time or a form of dilemmas involving ongoing courses of action.
- The VLITP management team comprising situations where a simple withdrawal might not be an obvious solution.
- The VLITP management team must come up with a real choice in deciding whether to persist or withdraw in the wake of mounting evidence against the VLITP viability.
- Refusing to recognize unambiguous feedback from previous decisions made during early stages of the VLITP implementation.

What Encourages VLITP Escalation?

Although a number of badly managed VLITPs are eventually terminated or significantly redirected, evidence suggests that a number of these projects are allowed to continue for too long before appropriate action is taken. Together with the often-visible monetary costs associated with these VLITPs, other more intangible costs include the following:

- Failure to solve real business problems for which the system was intended.
- An opportunity cost associated with spending valuable resources on a failing VLITP when these resources could have been put to alternative initiatives by the host organizations.

These three theoretical perspectives that may suitably explain the phenomenon of VLITP escalation: self-justification theory, prospect theory, and agency theory.

- *Self-justification theory* describes escalation as an attempt by VLITP managers to rationalize their previous behavior against a perceived error in judgment (Staw, 1981).
- *Prospect theory* describes escalation behavior as VLITP mangers acting contrary to the invariance criterion of rational choice (Tyler and Steensma, 1995).
- *Agency theory* views escalation behavior as the agent pursuing a course of action that is irrational from the principal's perspective (Eisenhardt, 1989).

Case II suggests that escalation is a complex phenomenon that may be influenced by many different factors. It agrees with the findings of by Keil and colleagues

(2000) of Information Systems Audit and Control Association (ISACA) members for the following purpose:

- Gather information concerning the frequency of IS project escalation (the frequency of this problem in practice).
- Identify major factors that may cause IS project escalation to occur.

Their startling results proclaimed that escalations occurrences of IS projects to be in 30-40% and that escalate are rarely completed and implemented successfully. Their findings suggested that escalation might have been caused by a combination of project management as well as psychological, social, and organizational factors. Our cases in this book also agrees with their suggestions that VLITP are more prone to escalation when they involve a large potential payoff, when the project requires a long-term investment in order to receive any substantial gain, and when setbacks are perceived as temporary problems that can be overcome.

Case I demonstrates a psychological factors, where the VLITP director became convinced that things do not look so bad after all, or that persistence will eventually pay off. This factor was promoted by the director's previous experience in handling similar projects—though at a much small scale—leading to the degree to which the VLITP director felt personally responsible for the outcome of the project. While VLITPs are more prone to escalation when there is a previous history of success as well as a high level of personal responsibility, the human tendency to 'throw good money after bad' in an effort to turn around a failing project can be another important factor (Bowen, 1987; Bunker et al, 2007). Finally, managers may also engage in a type of self-justification behavior in which they commit additional resources because to do otherwise would be tantamount to admitting that their earlier decisions were incorrect.

In Case II we see social factors being the major catalyst to escalation. These factors include competitive rivalry that may exist between different groups in an organization, the need for external invested over time. Thus, escalation can be thought of as continued commitment, justification, and social norms that dictate what is considered as socially acceptable behavior (Keil, 1995). Similarly Case I demonstrate that escalation is more likely to occur when external stakeholders have been led to believe that the project could be successful in the face of adversity.

All three cases demonstrate organizational factors—involving the structural and political environment that surrounds a project—thus, encouraging VLITP escalation. These factors can include political support for the project from powerful managers, the extent to which a manager's power or political fortunes are tied to the success of the project, and the degree to which the project becomes institutionalized with the goals and values of the organization (Keil et al, 2000).

VLITP Escalation Model

The managers of VLITP become locked in an escalation phase by means of making a series of erroneous decision. The proposed VLITP escalation model builds on theme in common project management literature, giving a promising theoretical base for analyzing and to some extent explaining what to avoid during the implementation of VLITP. Figure 11.1 shows several antecedents of VLITP escalation, including these determinants of escalation phase: project determinants, psychological determinants, social determinants, and organizational determinants:

- *VLITP determinants* are the objective attributes of a VLITP, its benefits as well as costs. A VLITP is quite often continued with high commitment because it is perceived as a long-term investment, expected to have a large payoff, and the host organization anticipates a long-term payoff structure. Such high commitment to a VLITP is also a result of closing costs being high and low salvage value anticipated.
- *Psychological determinants* cause VLITP managers to take an optimistic view. It also explains the VLITP managers' unwillingness to admit that an earlier decision could have been wrong. As a part of this determinant is Self justification theory and prospect theory—where VLITP managers exhibit risk averse or risk seeking behavior depending on how a problem or decision situation is framed—are psychological theories explaining VLITP escalation. Another psychological determinant is the human irrational economic behavior to "throw good money after bad" in an attempt to positively affect the future outcome of a failing VLITP. The champion of a VLITP did not only initiate

Figure 11.1. VLITP escalation framework

Theories/ Determinants	*Self-justification*	*Prospect*	*Agency*
Project	Case I		Case III
Psychological		Case II	
Social	Case I Case III	Case II	Case 1
Organizational	All Cases	Case II	All Cases

the VLITP but is also held responsible for its success or failure and would therefore be more likely to continue advocating its cause. Moreover, the more the general public gets to hear about the negative outcomes of the VLITP, the less likely it is for the VLITP managers or champion to change their original decisions.

- *Social determinants* in the case of VLIPT is very much associated with social comparison theory which proposes that VLITP mangers evaluate their attitudes and behavior in relation to others on the VLITP team and are likely to regard the behavior of others as a model for their own behavior (Keil et al, 1995). Such evaluation process is most apt to occur when VLITP managers are uncertain about the appropriateness of their own attitudes or behavior. Social determinants also involve a sub-project team's relation to another sub-project members. A successful effort by one sub-project team may influence another sub-project team to attempt the same approach. More important to this theory is that behavior of VLITP members are vitally affected by their relative power position on the VLITP.
- *Organizational determinants* affect the structural and political environment of a VLITP, including top management support, administrative inertia, power and inter-organizational interaction. Case I shows that VLITPs are more prone to escalate when there is strong political support and when projects become institutionalized. Institutionalization occurs when a VLITP is tied integrally to the values and purposes of the host organization (Guah and Currie, 2007). In such cases, many actions are taken for granted because the objectives of the VLITP are so deeply imbedded in the subculture or norms of the host organization (Bowen, 1987; Kakabadse and Kakabadse, 2002). It then becomes quite clear that long-standing operations and core product or services become impossible to consider for discontinuation because they are very much identified with the host organization VLITP initiative.

CHAPTER SUMMARY

Outsourcing can result in significant benefits for both the host organization and the outsourcing vendor. However, an appropriate approach needs to be followed to help ensure a successful relationship. This chapter has detailed an approach that begins with a baseline to fully understand the current environment and develop a fair VLITP contract with the outsourcing vendor. Management processes need to be reviewed and modified to help manage the relationship between the host organization and the outsourcing vendor. By using a form of metrics and auditing, the results quite often becomes necessary to manage vendor performance and the ongoing relationship.

The concept of escalation refers to the human tendency to adhere to a course of action even in the face of negative information concerning the viability of that course of action. The chapter has defined the circumstances under which a VLITP can be consider to be escalated as when decision makers at the host organization become over-committed to previous decisions about a VLITP, and invest more resources in a failing project. There are several antecedents of VLITP escalation, including: project determinants, psychological determinants, social determinants, and organizational determinants:

REFERENCES

Agrawal, M., Kishore, R., & Rao, H. (2006). Market reactions to E-business outsourcing announcements: An event study, *Information and Management, 43*(7), 861-873.

Ardagna, D., & Francalanci, C. (2005). A cost-oriented approach for the design of IT architectures. *Journal Information Technollogy, 20*(1), 32-51.

Bowen, M. G. (1987). The escalation phenomenon reconsidered: Decision dilemmas or decision errors? *Academy of Management Review, 12*(1), 52–66.

Brockner, J. (1992). The escalation of commitment to a failing course of action: Toward theoretical progress. *Academy of Management Review, 17*(1), 39–61.

Bunker, D., Kautz, K-H., & Thanh Nguyen, A-L. (2007). Role of value compatibility in IT adoption *Journal of Information Technology, 22*, 69–78.

Chau, P. Y. K. (1997). Reexamining a model for evaluating information center success using a structural equation modeling approach. *Decision Sciences, 28*(2), 309-334.

Di Romualdo, V. Gurbaxani (1998). Strategic intent for IT outsourcing. *Sloan Management Review*, 67-80.

Dos Santos, B. (1991). Justifying investments in new information technologies. *Journal of Management Information Systems*, *7*(4), 71–89.

Earl, M. J. (1996). The risks of outsourcing IT. *Sloan Management Review*, 26-32.

Eisenhardt, K. M. (1989). Agency theory: an assessment and review. *Academy of Management Review, 14*(1), 57-74.

Feeny, D. F., & Willcocks, L. P. (1989). Core IS capabilities for exploiting information technology. *Sloan Management Review*, 9-21.

Graham, M., & Cable, D. M. (2001). Consideration of the incomplete block design for policy capturing research. *Organizational Research Methods*, *4*(1), 25–45.

Guah, M. W. (2008). Changing Healthcare Institutions with Large Information Technology Project. *Journal of Information Technology Research, 1*(1), 14-26.

Johnson, J. (1995). Chaos: The Dollar Drain of IT Project Failures. *Application Development Trends* 2(1), 41-47.

Kakabadse, N., & Kakabadse, A. (2002). Software as a Service via Application Service providers (ASPs) Model of Sourcing: An Exploratory Study. *Journal of Information Technology Cases and Applications, 4*(2), 26-44.

Keil, M., & Montealegre, R. (2000). Cutting your losses: Extricating your organization when a big project goes awry. *Sloan Management Review*, *41*(3), 55–68.

Keil, M. (1995). Pulling the Plug: Software Project Management and the Problem of Project Escalation. *MIS Quarterly*, *19*(4), 421-447.

Keil, M., Mann, J., & Rai, A. (2000). Why software projects escalate: An empirical analysis and test of four theoretical models. *MIS Quarterly*, *24*(4), 631–664.

Keil, M., Mixon, R., Saarinen, T., & Tuunainen, V. (1995). Understanding Runaway Information Technology Projects: Results from an International Research Program Based on Escalation Theory. *Journal of Management Information Systems, 11*(3), 67-87.

Lacity, M. C., & Willcocks, L.P. (2006). Transforming back offices through outsourcing: Approaches and lessons. In Leslie P. Willcocks and Mary C. Lacity Global Sourcing of Business and IT Services, Palgrave Macmillan, pp. 97-113.

Lee, J., & Kim, Y. (1999). Effect of partnership quality on IS outsourcing success: conceptual framework and empirical validation. *Journal of Management Information Systems, 15*(4), 29-61.

Lee, M. K. O. (1996). IT outsourcing contracts: practical issues for management. *Industrial Management and Data Systems, 96*(1), 15–20

Mahnke, V., Wareham, J., & Bjorn-Andersen, N. (2008). Offshore middlemen: Transnational intermediation in technology sourcing. *Journal of Information Technology, 23*, 18–30.

McFarlan, F. W., & Nolan, R. L. (1995). How to management an IT outsourcing alliance, Sloan Management Review, pp. 9-23.

Ranganathan, C., Krishnan, P., & Glickman, R. (2007). Crafting and executing an offshore IT sourcing strategy: GlobShop's experience *Journal of Information Technology, 22*, 440–450.

Gonzalez, R., Gasco, J., & Llopis, J. (2006). Information systems outsourcing: A literature analysis. *Information and Management, 43*(7), 821-834.

Shi, Z., Kunnathur, A. S., & Ragu-Nathan, T. S. (2005) IS Outsourcing management competence dimensions: Instrument development and relationship exploration. *Information and Management, 42*(6), 901 – 919.

Staw, B. M. (1981). The escalation of commitment to a course of action. *Academy of Management Review, 6*(4), 577–587.

Staw, B. M., & Ross, J. (1987a). Behavior in escalation situations: Antecedents, prototypes, and solutions. In B. M. Staw and L. L. Cummings (Eds.), *Research*

Tyler, B., & Steensma, H. (1995). Evaluating technological collaborative opportunities: A cognitive modeling perspective. *Strategic Management Journal, 16*(Special Issue), 43–70.

Wallace, L., Keil, M., & Rai, A. (2004). How software project risk affects project performance: An investigation of the dimensions of risk and an exploratory model. *Decision Sciences, 35*(2), 289–322.

Willcocks, L., Feeny, D., & Lacity, M. (2004). IT and business process outsourcing: the knowledge potential. *Information Systems Management 21*(3), 7-15.

Zardkoohi, A. (2004). Do real options lead to escalation of commitment? *Academy of Management Review, 29*(1), 111–119.

Chapter XII
VLITP Management Framework

ABSTRACT

The traditional way to achieve the automatic execution of project management processes is to develop or purchase an application that executes the steps required. However, in practice, these applications only execute a small part of the overall process. Execution of a complete VLITP management process can also be achieved by using a framework of software with human interfaces, where needed when applications are not able to execute the management process automatically. Certain aspects of VLITP management process can only be accomplished with human interventions. Due to the complexity of managing VLITP, changing the scope is costly and an overview of the VLITP and their state is difficult to obtain. In order to effectively deliver its objectives, VLITP often requires that the underlying activities be constructed according to the principles of a framework. Thus, it is often difficult to make a suite of existing legacy systems fit with a VLITP. The commercial project management market has focused on specific process models to reduce the complexity of VLITP management This chapter presents a framework that provides the basis for identifying and describing generally accepted principles and practices of VLITP. The purpose is to guide the project management team of VLITP in the execution of one of the most challenging jobs in business today.

INTRODUCTION

There are numerous reasons why IT projects often never come to fruition. This book has detailed the issues involved with implementing VLITPs. The author anticipates that by understanding what leads to failure of VLITPs, readers would understand what is required not only to minimize the risks but also to dramatically improve chances of completing on time and under budget. By delegating a full chapter to methodological framework and approach for VLITP, the book demonstrates that business focus must lead as technology supports the efforts of meeting organizational goals of the business. The framework that follows is specifically modular and high-level to fit with the focus of this book—managing all types of VLITPs—which is combining business process management, various best practices for project management and using techniques from several software development frameworks.

The book shows how a VLITP is often an essential factor in the success or failure of the host organization, and forms the focus of the vital relations, which a business maintains with its external environment. Because technological competence of an organization is equal to the sum of the technological competencies of its various units, VLITPs are at the center of the definition of businesses in a 21st century organization (Lane et al, 2001). This requires that VLITPs occupy an important place in the organizational imagery and giving rise to important symbols (Currie and Guah, 2006). Thus, VLITPs are frequently at the very core of management strategy, corporate culture and organization identity. Unlike those who see VLITP as an extraneous variable imposed on the organization, this book shows how strategic management of innovative initiatives (like VLITP implementation) is an essential component of business policy in the 21st century. Thus, it must be understood and actively practiced by top management of multinational organizations.

Sufficient space has also been given to the imagery issue about host organization, the ideal role of employees and the importance of properly managing the implementation process of VLITPs. Also emphasized in this book are the issues of power structure within host organizations which can also help to cement a set of shared values and practices around the objectives of VLITPs as well as serve to complement and enrich the implementing process.

The next section goes through a proposed framework for implementing VLITPs. It takes into account the various topics explained in this book and concludes with the need to manage VLITP as a major change process in the host organization. Each stage of the VLITP is demonstrated by both the business and technological perspective of the project management team (see Figure 12.1).

Preparation and Decision Making

The first stage in VLITPs is the preparation and decision making for the project. Here, there are certain issues that must be addressed before the project can proceed to the next stage. The first thing to ensure success at this stage is that business takes the lead. The main goal of this stage is to define what the scope of the project is should be. Critical to this is that both business and IT managers must understand and support the goals of a VLITP, document must be in place with agreement concerning the scope and goals of the VLITP before moving on to the next stage of the project.

Business Perspective

This prospective demand that project managers ensue the following needs are addressed before moving on to the next stage:

- Determine strategic and project goals
- Determine the key success factors for the project
- Determine the scope of the project
- Determine the functional requirements
- Create support for the project
- Writing a project plan that focus on the tasks, resources, deliverables, budget and a timetable for each activity/task.

IT Perspective

From a technical IT perspective, the following issues need to be addressed:

- Determine the technical requirements
- Write a project plan that focuses on the tasks, resources, deliverables, budget and a timetable for each activity/task.

Analysis

At this stage an analysis is made of the current situation and the risks, processes and changes are identified. The main goals are to analyze the current situation, identify the demands for the new situation and identify all the risks and changes that need to be made (Neale et al, 2007). Business and IT need to understand the current situation and changes that are to be made and be in agreement about these changes and the way these changes are implemented before proceeding with the next stage. In this stage the alignment of business and IT goals for VLITP is critical before diving into the actual building and implementation of the project deliverables.

Business Perspective

Based on the business requirements the following need to be addressed during this stage:

- Map the current situation within the host organization
- Categorize and identify the processes to be affected by the VLITP and subsequent risks involved
- Make an analysis of the changes to be introduced by the VLITP
- Determine the VLITP output required
- Pay attention to the Internal Control aspects for the processes that would be impacted by the VLITP
- Analyze and optimize such processes
- Simulate such processes.

IT Perspective

The technological perspective considers the following issues to need addressing during this stage:

- Determine the technical deliverables to be provided by the VLITP
- Investigate Internal Control aspects for the processes that are most likely to be automated
- Make an "as-is to-be" analysis for the VLITP.

Design and Realization

This stage mostly focuses on the design and delivery of a proper blueprint for the actual system and processes to be implemented from the VLITP. This stage also considers the need to plan for maintenance after designing the system. That dictates any change must be accurately documented after approval within this stage before it is realized. In instances where complications are found in this stage, all relevant parties involved in the design process must ensure the overall impact on the VLITP is kept to the minimum. Thus, changes and decisions relating to IT functionalities are made before those of business considering all decisions on the changes in IT functionality have to be endorsed by business (Kim and Kwahk, 2007).

Business Perspective

This perspective requires that the following need be addressed here:

- Making a blueprint for new processes, in which the risks and controls for all new processes are implemented.
- Testing functionalities by users to see if the blueprint will be effective in practice.
- The designing of functional and operational management. Doing this within VLITP provides for processes and activities that are necessary for providing proper information structure within the host organization.
- Involving potential users and employees at this early stage usually help in identifying specific needs and getting valuable inputs for system design purposes.

IT Perspective

From technical IT perspective, the following issues need to be addressed here:

- Defining the modules of the LVITP.
- Define a list of technical specifications based on the blueprint (which is made from the business perspective) and the technical requirements which were defined in the previous stage, taking into account the valuable users input.
- Design the IT system architecture in terms of hardware, software, database, and user interface.
- Testing of the IT system to make sure that requirements are met.
- The design for the technical and application management within the organization. This can be done with ASL or ITIL to provide guidelines for processes and activities that are necessary for a proper maintenance and control for the VLITP.

Implementation

This stage is where the necessary changes and resulting systems are implemented within the organization. The main goals here are that at the time the VLITP "Goes Live", the users would be adequately trained and familiar with the new system to ensure the smooth continuation of the business operation. It should also ensure that the maintenance organization is in place to support the new system.

Business Perspective

From business perspective the following need to be addressed during this stage:

- Writing the training material and working procedures for the VLITP.
- Coordinating the user training process.
- Giving procedural training to all users and maintenance organization on the VLITP.

IT Perspective

Technologies issue here involves the following:

- Writing manuals for VLITP deliverables.
- Coordinating the training of the IT department.
- Construction, testing and installation of the IT system.
- Giving system training for the users and maintenance organization of the IT system.
- Go Live

Maintenance and Support

The final stage of any VLITP emphasise the primary goals of recording lessons learned during the project as well as looking for continuous improvements.

Business Perspective

From the business point of view the following need to be addressed during this stage:

- Auditing.
- Running the functional and operational management.
- Monitoring, based on the critical success factors, Key Performance Indicators and with the help of a Balanced Score Card.
- Analysis the performance on the operational level.

IT Perspective

The technological issues here are as follows:

Figure 12.1. VLITP methodological framework

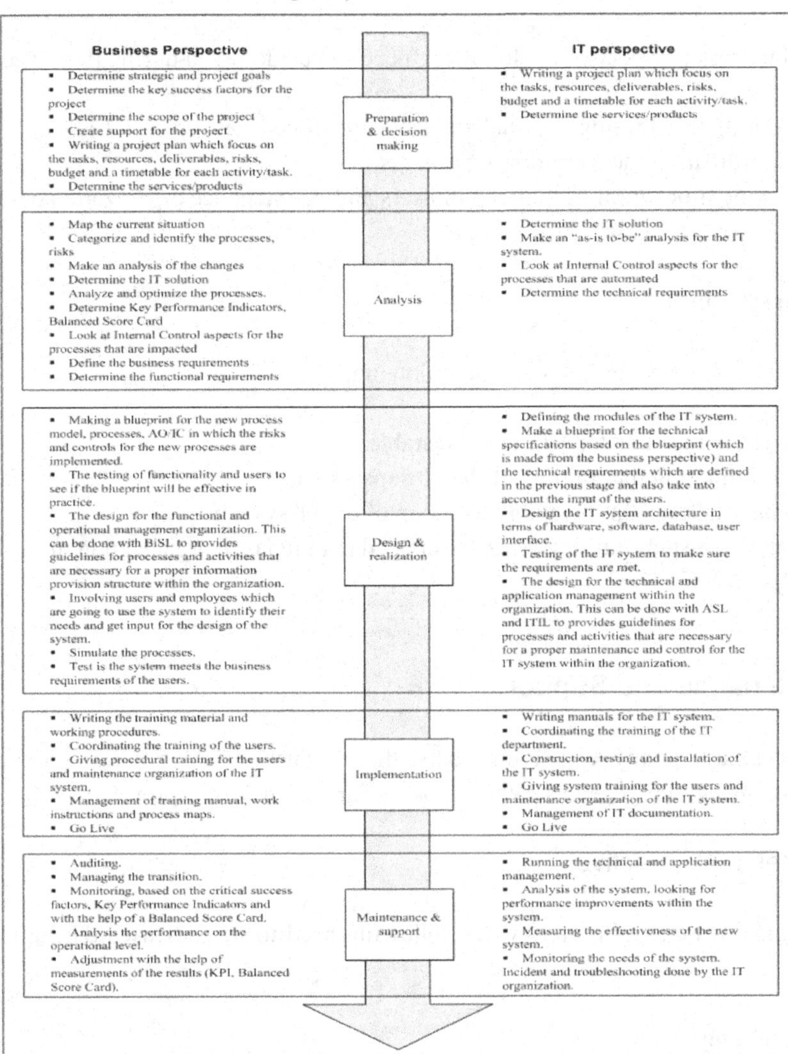

- Running the technical and application management.
- Analysis of the system, looking for performance improvements within the system.
- Measuring the effectiveness of the new system.
- Monitoring the needs of the system.

This methodological framework (see Figure 12.1) for VLITP is a high level description of how a VLITP should be carried out. As a tool, it supports the proper management and objectivity of a VLITP. The author does not suggest it as an absolute way of carrying out a VLITP. Rather as a means of taking into account the risks involved.

Leadership

VLITP leaders have the skills to set strategy, think laterally and lead by example. A good VLITP manager should be able to see where the project is going and therefore visualize see how to get there. Leading a VLITP means being creative which is not something one learns but rather an attitude one maintains. Leading a management team for a VLITP requires trust and integrity that demonstrates what one stands for when performing the role leadership plays in effectively executing VLITPs take charge, instinct, and innovative thought (Guah, 2008). It has been made cleared, all through this book, that the management of VLITPs is more distinguished from the shadows of narrow focus on schedules and meetings to achieve the full potential. The work of VLITP management team is not just about the end game of on-time and under budget but also getting the true message of what the projects stands for to the corporate executive of the host organization (Feldman, 2004). Managers of VLITPs are expected to understand their own strength and the strategic implications of their actions.

In order to ensure VLITPs succeed in delivering intended objectives, the management team should keep the following factors in mind:

- Avoid the temptation to cut corners, financially or methodologically. It usually results in project failure or inadequate objectives that do not meet the users' needs.
- Audit each major deliverable and sub-projects along the way for accuracy and correctness.
- Carefully monitor board and top management of host organization continuous support and interest for the project. Make sure that they are aware of the progress of the project management team.
- Recruit and maintain in position the correct technical lead for VLITPs.

The book has also warned of taking 'free lunch' in managing VLITP. If the host organization insists on keeping costs low and hurrying the project along, then quality will be low or the risk of failure will be high no matter how well the project is managed.

The four characteristics of bureaucracy—specialization, hierarchy of authority, systems of rules, and impersonality—are continuously being used within VLITP because they are the most efficient method of managing project yet discovered. Management teams of VLITPs often experience both advantages and disadvantages when dealing with bureaucracy. While it may suit certain temperaments better than others, bureaucracy brings with it the problem of balance within VLITPs. The project managers are faced with inescapable problems that can be ameliorated. Some of them have to do with:

- The need to have rules that are impersonal but not inhuman.
- Fair yet taking the individual into account.
- A structure that is not too rigid to adjust to change.
- Be loyal to further the host organization's objectives without developing too many superstars on the project.

Project Champions

While the success of a VLITP may be attributed to a wide range of variables, the book has also demonstrated that some individuals contribute much more to VLITP success than others. They do this by promoting the project decisively and enthusiastically throughout the stages of planning, implementation and maintenance. Such individual identified in this book as champions, leaders and change agents, can strongly impact the ultimate success of a VLITP

These champions are identified and differentiated by how they generate and control the strategy needed to successfully implement the VLITP. They quite often not only possess power and prestige in the host organization, but also control sufficient funds and have the authority to facilitate the implementation of VLITP. A potential VLITP champion would normally be able to recognise the outcome of VLITP as having significant potential to the host organization. The individual be in the position to make personally comments to the particular VLITP without any conflict of interest. This would facilitate the role of advocating vigorously for the VLITP as well as generating the requisite support from other people in the organization.

Risk

Risk of failure is present at all stages of a VLITP, regardless of the nature of the industry, or the environment in which it is undertaken. This book has expressed the need for doing a formal p*roject risk analysis and as part of VLITP management strategy. This exercise should* focus on all issues that affect various subsets in the VLITP. It should not only focus on project specific issues, but also address how the risk management processes at subproject level connects to the overall objective and the business goals of the host organization (Brown and Jones, 1998; Gallivan, 2001). The procedure should be available to everyone that has a role within the VLITP including the various managers of each sub-projects, the host organization, contractor, sub-contractor, consultant or service providers. The following were among the *risk analysis and management procedures* observed in the projects during the research for this book:

- Incorporates the latest practices and approaches to risk management in projects.
- Coverage of project risk in its broadest sense, as well as individual risk events.
- The use of risk management to address opportunities (uncertain events with a positive effect on the project's objectives).
- A comprehensive description of the tools and techniques required for that project.
- Relevant material on the human factors, organisational issues and the requirements of corporate governance within the project.
- One or more sections on behavioural issues within the project.

Mismanaging VLITP information could affect the achievement of a VLITP's objectives. This can lead to a course of actions that misinterprets the actual plans and how the management team hope to attain the project goals. Proper planning of VLITP prevents the host organization from putting its investment at the mercy of the project management "whims" as well as other external pressures, which often lead to VLITPs loosing sight of their objectives. This is an aggregate of the significantly low level of forward planning by businesses against the often far too rigid planning in the public sector. The private sector businesses are content with carrying on operations virtually on a day-to-day basis. On the other hand public service organizations takes very little account of likely and possible political and economic changes.

The risks management process of VLITP begins with risks identification wherein a VLITP is assessed for potential risks that could threaten the project process itself or any of the anticipated outcomes. Considering identification is only a starting point, identified probable risks must then be assessed in order to determine the level of impact (Schwarz and Hirschheim, 2003). The estimated impact could be negative or serious which may raise other questions whether or not the probable risk warrants additional action at this stage of the VLITP.

Contract

Most VLITPs are implemented by IT service providers who are contracted to do so. From their point of view, the basis of a successful VLITP is a very good contract that has been well thought through. The contract must clearly define what work is to be done, by whom, when and to what standards. Within every good VLITP certain form of contract strategy is applied. The strategy enables the creation of contracts which record the obligations of the signatories, allocates risk in the most appropriate manner and is seen as fair by all parties. A well-written VLITP contract is used to resolve difficulties as painlessly as possible, especially when anticipated and unanticipated problems arise, or when variations are necessary, or when schedules overrun during the implementation of a VLITP. The most important parts to the contract document include:

- A strategy for procurement and sub-contracting sub-projects in the VLITP.
- Contractor selection process.
- The contracts scope.
- The risks level.
- Terms of payment.
- Planning strategy.
- Contract management policies.
- Dealing with unanticipated problems during the implementation of very large IT projects.

Managing Benefits

Even when the objectives of VLITPs are delivered successfully the question still remains if the project team is delivering the expected business benefits. By applying rigorous benefits management strategy, expected benefits can increase to about 85%, with an additional project of program overhead that should not exceed 5%. Good benefits management during the implementation of VLITP is characterized by a set of principles—though such principles can be difficult to apply in practice. The management team of a VLITP must follow a step-by-step approach for a chance to ensure that the project doesn't fall into the trap experienced by many IT projects today. Many VLITPs have successfully implemented systems resulting to increase in business volumes and profits after the change is implemented (Taylor and Todd, 1995). The surprise is often that along such increase comes dramatic falls in sales to previous levels. Further investigations show that the uplift was due to market conditions rather than the new system. In certain instances customers have seen little improvement in service levels because a large part of the sales force had not changed their ways of working in line with the improvement systems (Lane et al, 2001). By the time these issues had been addressed, competitors had had time to catch up with their innovations and the project value was lost. This is an example of how organizations can successfully complete a systemic change by implementing VLITPs but then fail to achieve the anticipated business benefits.

Financial Accountability

VLITPs are expensive to implement. They can easily run over budget, especially on projects that last for extended periods. Both in USA and Europe, experienced IT personnel are commanding larger salaries and more extensive perks (such as cash hiring bonuses) every year. This environment leads many of the most talented employees to change jobs more often, and higher turnover rates lead to increased costs (for training, recruitment, and higher, more competitive salaries). Delays in completion also lead to cost overruns.

Some management teams try to overcome cost overruns by reducing or even eliminating certain good designs and certain aspects of project management skills. This strategy to control VLITP finances encourages the reuse of standard components, whether commercial products or items created by project staff, and can dramatically reduce costs and minimize the risk of cost overruns. Reducing staff turnover is also key, whether through excellent leadership and management practices, competitive compensation packages, or career advancement opportunities.

For any of these strategies to succeed the project management team needs to understand the issues around VLITP accounting. While VLITPs can be implemented to deliver strong reporting both on individual units as well as across the organisation as a whole, reporting the total cost of VLITPs including specific over-spends or under-spends would help the project management team's effort to focus on improving overall effectiveness and efficiency in the host organization. Their final deliverable was expected to drive improved efficiency, reduced costs and greater return on the investment of the host organization. Though members of a VLITP management team may not be accountants, they are expected to give accurate reports on expenses and costs, as well as delivering their work more efficiently on every new project. Such reasoning brings additional requirements to the skill set of the management team of a VLITP. This book has identified some of the skills needed to manage VLITPs, including:

- Cost accounting knowledge in project management.
- The ability to guide the project expenditures by providing cost information speedily.

- Enabling the development of internal efficiency in the VLITP to achieve a greater return of investment for the host organization.
- To educate the host organization about the salient issues of any change process.
- Continues updates on VLITP financial requirements during implementation.

During the implementation of VLITPs, unexpected questions and problems arise, and where the outcomes of the project hang in the balance, there could be one or more concern uncovered. Such concerns can pop up at any time during a VLITP, and must be dealt with quickly, without the benefit of pre-defined solutions. They may include circumstances surrounding project deliverable, or even elements of the project process—such as schedule, scope, budget or other project parameter.

SMOOTH TRANSITIONS: PLANNING PROJECT CLOSURE

All VLITPs must come to an end, one way or another. While a large number do come to an untimely end—through cancellation—most reach their planned conclusion. This 'conclusion' is a part of the VLITP management process, usually referred to as project closure. Closure is often considered a sign of success and achievement, and should be treated as such, ensuring that successful VLITPs go out with a bang, and not a whimper. Only in rare cases does the performance of new product resulting from VLITPs be so poor that the host organization decides it may prove less expensive to discard the entire project and start from scratch than attempting to repair the defects.

SIZING PROJECTS

Project sizing is a 'must' for VLITP implementation strategy, ensuring that all plans and activities are relevant, and that resources are properly used and allocated. This is particularly important in a VLITP environment, where simultaneous sub-projects must compete for both financial and human resources. These methodologies for sub-projects would never fit the entire large-scale project, and most smaller projects would easily be overtaken by the weight of overly detailed procedures and practices.

Delays in Completion

Missing dates are the plague of implementing all projects in the IT industry, with delays of six months to a year common. Most delays are the results of changes to the original objectives of VLITPs, changes to the underlying architecture of the system, and poor estimating at the start of the project (Westrup, 1998). They can also be a result of the departure of one or more key members of the project implementation team—a new member filling the gap quite often need to spend extra time getting up to speed with the project activities.

A VLITP schedule reflects the work to be done as outlined in the specification. Specification is meant to drive the schedule, not the other way around. When a lot of late changes are needed to a particular specification, that usually indicates that not much time was spent on the specification and what looks like 'changes' are really the specification being created. Specifications are often compromised in order to get systems developers working on the outcomes of a VLITP. The specification serves as an implementation guide for development and test. At a minimum, the development iteration at hand should have a clear specification to guide the work of the project teams members.

The fact that these issues occur is no surprise but what is often disappointing is that the management team fails to inform relevant parties about them in time for the host organization—and their customers—to take the necessary precautions. That is probably because they fail to carefully monitor progress on these issues within the project. When interim deadlines are missed, VLITP management team should immediately find out what is going wrong and take the appropriate steps to correct the problems, and inform the relevant parties within the host organization. This removes any obstacles that are impeding the project, by allowing the management team to allocate more staff members to it, or change the assignments of your existing staff. It is important, however, to be careful not to cause additional problems when modifying the project timetable, or it might cause even bigger problems later in the project. The important issue is to act quickly to dissimulation of project information. The longer a project proceeds with delays, the harder it is to get back on schedule.

Insufficient Testing

Once a VLITP has been completed it must be tested thoroughly before being deployed throughout the host organization. That way, the defects can be resolved before the host organization begins relying on the project for its business. However, many staff of VLITPs often scrimped on testing. Not only is it an expensive initiative to create a proper test plan, it is also very expensive and time consuming to conduct the actual testing. It also takes time, and if a VLITP is close to its due date, the project staff will be under a lot of pressure to deliver. Thorough testing is usually one of the first casualties of VLITP facing tight deadlines or budgetary pressures.

When the outcomes of VLITPs are not tested properly, they will usually have serious defects or will not perform adequately when placed under a full operating load. These problems will wreak havoc in a business environment, often affecting the image of the host organization in the public. VLITP management teams should envisage this problem before announcing closure.

Project Reviews

One of the most difficult tasks for the management team is to complete a project review when a VLITP had not gone well. There is however, a need to carry out a review, regardless of the outcome of a VLITP. Quite often stakeholders in the host organization are looking for recommendations to get a VLITP back on track after a previous fails. When projects get into trouble it is often for one (or more) of the following reasons:

- Lack of a clear vision or business case
- We don't know what we don't know
- Scheduling without specs and late changes to specs
- Schedules based on desire, not reality
- Inadequate resources
- Cross-product dependencies
- Vendor dependencies

Lack of a Clear Vision

Publicity alone is not sufficient for VLITP success. The business case of VLITPs should reflect compelling, understandable and achievable business goals. A number of IT projects have begun on the premise of nothing more than a verbal request, with little or no analysis behind it—another evidence for the present rate of IT project failure. Unlike a standard IT project the vision of VLITP should articulate why the project is being undertaken, explicitly note what the host organization expects to achieve, rationale for any deadline, which categories of customers are expected to benefits from various deliverables, and what would or would not be part of the project scope. By clearly articulating these foundational materials, the VLITP can avoid being under a cloud of uncertainty from the initial stages.

The Unknown

This problem of 'unknown' occurrences during VLITP implementation is often a result of the management team failure to admit that there are key areas of IT knowledge/skills, business climate factors, and other important items, about which they actually lack experience. It is usually in an effort to preserve their egos that the management team leads stakeholders to believe they know what they are talking about (Brown and Jones, 1998). One way to accommodate uncertainties when implementing VLITP is to adopt a feature-driven development (FDD) approach. By producing a very preliminary functional specification at the beginning, with the most important features being detailed and developed in the earlier iterations of development, FDD allows the management team to adapt as better information becomes available. FDD can be combined with the use of risk management and schedule buffer to allow management teams to make mid-course corrections while managing stakeholder expectations.

Schedules Based on Desire

Too often when a VLITP begins, the only thing known is the deadline. The definition of the rest of what is to be done is pretty much content free. While there is nothing inherently wrong with an aggressive release date, it must also be realistic. The time to talk with relevant parties about unrealistic deadlines is after the team has had time to assess what can and cannot be accomplished (Argarwal and Prasad, 1998). And this means after some preliminary specification and schedule development has been done. It is much more helpful to tell relevant parties what they can have and by when than to commit to a fantasy project doomed to failure.

The host organization consists of individuals who may nor may not function well under extreme schedule pressure. One of the ways this manifests itself in host organizations is by assigning the same individuals to multiple priority-one projects of which the VLITP would be connected. Using a shared resource approach, instead of getting the most important project completed and out the door first, that project is doomed to be completed at the same time than the least important sub-project within VLITP.

Vendor Dependencies

Every member of VLITP management team would have had experience with vendors not delivering on time, not delivering what was specified, trying to push through technical leads at the initial meetings and even staffing the project with less experienced technical staff. By taking a closer look at the contracts of many VLITPs, it can be observed that they do not specifically state how review and acceptance authority regard the initial and subsequent vendor staff. Due diligence should be paid in time to finding qualified vendors based as what they spend in recruiting and hiring full-time staff. Too often, in the rush to bring a vendor on board, little time is spent putting the specifications into the contract and defining the terms of business relationship for VLITPs. To avoid later difficulties, management team should spend time specifying the deliverables, defining unambiguous acceptance criteria, determining how and when payment is to be made, stating review and acceptance authority over interim and the final outcome, and vendor requirements for correcting errors to the VLITP after release.

VLITP IMPACT FACTORS

The framework in Figure 12.2 represents various factors that usually lead to the prediction of a successful or unsuccessful VLITP. Figure 12.2 shows the need for project goals to fit the required tasks. Important conditions for the perfect match are attitudes and behavior within the project environment being suitable for an overall impact. Several determinants make up the assessment criteria for fitness for utilization of VLITP outcome (see Figure 12.2). The project fitness assessment is determined by the correspondence between the task and the people/organization characteristics, mediated by the ability of the individuals. The possible utilization

Figure 12.2. Determinants of VLITP impact factors

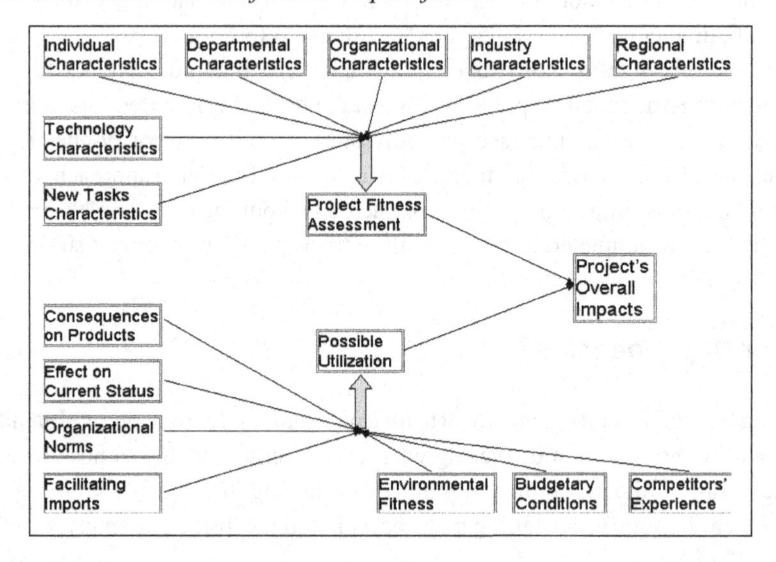

is determined by arbitrary factors about the products or services being provided by the host organization and the condition of the industry (see Figure 12.2). Together they suggest the necessary things to have right at the initial stages of a VLITP.

CHAPTER SUMMARY

The field of project management uses several different standards and best practices to increase the likelihood of overall success. Individual standards and practices may vary in complexity and application, but the goals are usually the same—to produce desired project results within the boundaries of time, costs and available resources. Any effective program for VLITP management standards and best practices must provide relevant steps and strategies to guide the selection, management and control of projects pending and underway. Every VLITP should begin with an approved requirements specification. But requirements rarely present themselves. Before VLITP requirements can be selected and approved, they must be collected, culled and defined.

The management of costs can be the most complicated and potentially tedious element of the VLITP management process. But, costs and expenses have to be controlled to ensure credibility and the overall fate of current and future projects. VLITP cost management involves cost factor identification, budget creation, and cost control through variance tracking.

Communication is a vital tool in the successful management of VLITPs therefore its proper usage should be taken lightly. It is with seriousness that the author of this book encourages all management teams, supervisors, and team leaders of every VLITP to utilise this tool with the adequate effort it deserves. The practice of very good communication during the implementation of VLITP involve:

- Believing that communication is king to all elements of the VLITP.
- Taking standard approaches of communication that really work.
- Communicating opportunities for education and training including both structured and unstructured curricula for those on sub-project teams.
- Vigorously communicate accountability that applies the constraint of a rigorous return on investment to all sub-projects with sufficient business cases for each.
- Clearly and promptly provision of portfolio and performance report to verify that the right job is being carried out.

When selling VLITPs, project champion frame them in relation to business results (using triple constraints involving cost, scope and time). Using this set of business language to manage the host organization's expectations of VLITPs objectives, frames VLITPs benefits as solving business problems and getting results. It also brings out the linkages between strategic management and VLITP implementation by explaining a clear fit between VLITPs and the organization. It emphasizes the expanded role of VLITPs as project portfolio, business risk management, resource management in accordance with the corporate direction of the host organization.

This book shares with the reader various ways of thinking through the implementation of VLITP—which includes successes, failures and some near catastrophes. The ultimate conclusion is that the best way to make a VLITP succeed is for VLITP managers to use their heads. The main concentration should be trying to avoid making the mistakes that would lead to VLITP failure. Surprisingly, the solution is most often simply common sense. Unfortunately the literature shows that common sense is often ignored when implementing VLITP. If that were not the case every VLITP would be totally successful where all problems would be avoided, the VLITP stages would move according to plan and to cost. All members of the VLITP team would efficiently get on with their jobs, and little problems would be solved before others not immediately involved see them. These can only result from engaging a totally successful VLITP manager who then becomes invisible. As a result the host organization would not need a manager for the next VLITP because ordinary business executives would feel the job is too easy for someone with enormous salary to be engaged.

REFERENCES

Argarwal, R., & Prasad, J. (1998) A Conceptual and Operational Definition of Personal Innovativeness in the Domain of Information Technology. *Information Systems Research, 9*(2), 204-215.

Brown, A. D., & Jones, M. R. (1998). Doomed to Failure: Narratives of Inevitability and Conspiracy in a Failed IS Project, Organization Studies, 19(1), 73-88.

Currie, W. L., & Guah, M. W. (2006) 'IT-Enabled Healthcare Delivery: The UK National Health Service'. *Information Systems Management*, Spring 2006, pp.7-22.

Feldman, M. S. (2004). Resources in Emerging Structures and Processes of Change. Organization Science,*15*(3), 295-309.

Gallivan, M. J. (2001) Striking a balance between trust and control in a virtual organization: A content analysis of open source software case studies. *Information Systems Journal, 11*, 277-304.

Guah, M. W. (2008). Changing Healthcare Institutions with Large Information Technology Projects. *Journal of Information Technology Research*, 1(1), 14-26.

Kim, H-W., & Kwahk, K-Y., (2007). Managing readiness in enterprise systems-driven organizational change. *Behaviour and Information Technology, 27*(1),.79–87.

Lane, P., Salk, J., & Lyles, M. (2001). Absorptive capacity, learning and performance in international joint ventures. *Strategic Management Journal, 22*, 1139-1161.

Neale, W. S., Smerek, R., Haaland, S., Denison, D. R., & Gillespie, M. A. (2007). Linking organizational culture and customer satisfaction: Results from two companies in different industries. *European Journal of Work and Organizational Psychology, 17*(1), 112–132.

Robbins, S. (2004). Is Governance. *Information Systems Management, 21*(4), 81-82.

Schwarz, A., & Hirschheim, R. (2003). An Extended Platform Logic Perspective of It Governance: Managing Perceptions and Activities of IT. *Journal of Strategic Information Systems, 12*(2),129-166.

Taylor S., & Todd, P. A. (1995). Understanding Information Technology Useage: A Test of Competing Models. *Journal of Information Systems Research*, 144-176.

Westrup, C. (1998). What's in information technology? *Implementation and Evaluation of Information Systems in Developing Countries*, C. Avgerou (ed.), Asian Institute of Technology, Bangkok, 77-91.

Zmud, R.W. (1984). An examination of push-pull theory applied to process innovation in knowledge work. *Management Science, 30*(6), 727-738.

Section IV
Case Studies

Chapter XIII
Case Study I: National Program for IT

ABSTRACT

The National Programme for Information Technology is the largest civil IT program worldwide at an estimated cost of £6.2 bn, US$ 10 billion, over a ten-year period. Launched in 2002, it provides an opportunity for the IT service industry to develop business models in the UK healthcare sector in which, historically, has seen low investment in IT services. Despite highly publicised large-scale IT outsourcing contracts, many IT vendors have been unable to fulfil the rigid terms and conditions of their contracts. The chapter provides current additional evidence to that found in the literature on emerging technologies in the health sector. It aims to investigate some of the issues that are associated with the implementation of the emerging technologies in the NHS and explores whether the implementation of the National Program, by the NHSIA, would bring value to patient care, and influence staff perception of IS.

INTRODUCTION

This chapter contains details about a real case of VLITP, based on a 4-years research involving more than 100 interviews in addition to other interactions (Guah, 2008). During the period 2000 to 2005, Connecting for Health (formerly called National Health Service Information Authority) was pushing through a highly complex procurement process—termed the National Program for Information Technology (this book refers to it as National Program). This project was estimated to costs the British taxpayers £6.3 billion for additional IT investment for the NHS over a ten years period (NHSIA, 2003). To understand how such a VLITP can affect the largest employer—and third largest in the world—the chapter also contains a narrative of IS strategy in the NHS, involving institutional theory and saturation theory, as part of a process of reforming healthcare delivery in the UK

The National Program utilised a comment emerging technology for most VLITP—the Application Service Provision (ASP) business model—to avoid similar fate as many previously implemented technological change in the NHS which resulted in failure (Collins, 2003; Currie and Guah, 2007; Eccles et al, 2002; Lauchlan, 2000; Marshall et al, 2003; McGinity, 2003). Researchers continue to question the nature, origins and applicability of various e-business models—including the ASP business model (Caldwell, 2002; Howcroft, 2001; Kraemer and Dedrick, 2002; Chatterjee et al, 2002). Improvement in the rate of successful IS project implementation may lie in concentrating on human and organisational aspects, rather than the technological ones (Scott-Morton, 1991; Davenport, 1993; Coombs and Hull, 1995; Bloomfield, 1997).

While this chapter DOES NOT attempt to interpret the political strategy for the present or any other UK government in dealing with the NHS, it provides a review of a particular reform initiative, using the lens of Institutionalisation of IS in healthcare delivery processes. This also supports the view that the human, political, social and cultural aspects of IS strategies need to be taken more seriously (Moad, 1993; Belmote and Murray, 1993; Mumford, 1995).

BACKGROUND TO THE NATIONAL PROGRAM

The NHS needed to adopt a revised model for IS strategy which takes into account additional complexities of remote sourcing via the model used in this VLITP. The driving force behind the NHS strategy is a political one, of providing socialised medicine to all, within a fixed national budget. That NHS is therefore cash limited at a total annual budget for 2004-05 of around £90 billion, about 8% of the national GDP (gross domestic product). Over the past decade, IS strategies for most organizations have been in a state of evolution toward a form of 'federal governance architecture' (Zmud, 1988). These are instances where the authority for the management of IS infrastructure is vested with a central IS department, but the authority for the management and use of IS application is vested with individual business units, or regional Trusts in the case of the NHS (see Figure 13.1). The NHS is a national institution, but is not a single corporation. It is funded mainly from general taxation and national insurance contributions. IS management in the NHS, therefore, involves a variety of coordination mechanisms. These include IS councils, IS steering councils, service level agreements, and charge-back internal accounting systems, as structural overlays to supplement the hybrid 'federal governance architecture'.

The availability of electronic health record (EHR) by spring 2005 served a catalytic role within the NHSIA strategy for the immediate future (DoH, 2002). In a seven-year strategic plan, the NHSIA presented a strategy based upon a vision of patient-centred data flowing freely across all areas of healthcare, to be accessed and utilized wherever it is needed (DoH, 1998). Several researchers have identified the availability of nation-wide EHR as a 'holy grail'. It has been promoted as a panacea to problems in healthcare information management. But researchers have also identified problematic aspects, including security and confidentiality issues that occur during data storage and transmission (Ferlie and Shortell, 2001; Jacklin et al, 2003; Laerum et al, 2001). Ferlie and Shortell (2001) also observed the following issues with a nation-wide EHR, which raise important problems:

- A lack of standardization (i.e. coding of medical conditions, vocabulary of medicine).
- A need to support different views on the structure.
- A need to reconcile different contents of a nation-wide EHR.
- A need to cater for different types of users with different requirements (physicians, patients, nurses, administrative staff, etc).

National Program is an initiative to implement VLITP by the NHSIA, born as a result of several plans to devise a workable IS strategy for the NHS (DoH, 1998; DoH, 2001; Wanless, 2002). The National Program was designed to connect the capabilities of modern IS to the delivery of the NHS Plan devised in 1998. The core of this VLITP is to take greater control of the specification, procurement, resource management, performance management and delivery of the information and IS agenda (NHSIA, 2003). In the forth quarter of 2003, the initiative began to award contracts to private service providers under the National Program. This is now the preferred nomenclature for the electronic care records that form the centrepiece of the NHS IS modernisation programme. Contracts have been signed with Local Service Providers (LSP) and National Application Service Providers (NASP) to deliver the NHS care records service (see Table 13.1).

Here are few objectives the National Program hopes to accomplish:

- Firstly, to have a series of tightly specified and priced framework contracts on a short-list (of about 5) of Prime Service Providers (see Table 13.1) who

Table 13.1. Service providers contracted as LSP or NASP

Primary Service Provision Model			
• NASP purchasing & integrating nationally • LSP delivering IT systems & services locally			
CONTRACTS (DEC-2004)	**SERVICE PROVISION**	**PROVIDER**	**LENGTH**
Care Records Service – NASP	National	British Telecom	10 years
Care Records Service – LSP	North East	Accenture	10 years
Care Records Service – LSP	Eastern	Accenture	10 years
Care Records Service – LSP	London	Capital Care Alliance (British Telecom)	10 years
Care Records Service – LSP	North West & West Midlands	CSC	10 years
Care Records Service – LSP	Southern	Fujitsu Alliance	10 years
National Network for the NHS (N3)	National	British Telecom	7 years
Choose and Book	National	Atos Origin	5 years
Contact (email)	National	Cable & Wireless	10 years

can work at local level. This could consist of integration and implementation partners, at a national level, to support all aspects of the National Program. Each PSP will have an aligned consortium of service providers and vendors for the ICRS element of the National Program, and will be mandated to work with the domain PSP for electronic booking, the infrastructure providers and healthcare providers. PSPs may not make their products exclusive or mandatory to their StHA.

- Secondly, to create priced packages of national services and applications that the PSPs and hospital administrators can together implement locally. This activity will include managing the creation of a single national Health Records Infrastructure Service (HRI) and other national services, to access and move health record information as required.
- Thirdly, to create Service Level Agreements (SLAs) for the national services and other services out of the scope of the PSP consortium (see Table 13.1), that the PSPs can work to in providing an integrated service to the StHA.
- Fourthly, to develop and maintain the national standards and specifications that all vendors must use. It is also anticipated to create the national business cases required for the DoH (required by Treasury) governance procedures. This shall support the local decision making business cases required at StHA level.
- Lastly, to procure, under national contracts, a backbone network infrastructure.

Table 13.2. Key issues in the national program

KEY PRINCIPLES	KEY COMPONENTS
* improve the patient experience * meet the best interests of the NHS * provide effective support for clinicians * represents best value for money for the taxpayers	* electronic booking of appointments (e-bookings) * electronic transfer of prescriptions (ETP) * an Integrated Care Records Service (ICRS) * an underpinning IS infrastructure with sufficient connectivity and broadband capacity to support the critical national applications and local systems.

Table 13.2 shows that key principles and key components for the National Program make a significant difference to patients and patient care.

The aim of NHSIA strategy for VLITP was taking greater control of the specification, procurement, resource management, performance management and delivery of healthcare information and IS agenda. To deliver the necessary applications, services and IS infrastructure required within a timeframe that was both sensible and ensured value-for-money, the VLITP management team needed to run a procurement process that was as rapid as the NHS culture allows (Mark and Scott, 1991). Such a vibrant procurement process brought to the NHS the benefits of maintaining management focus and better engagement of prospective suppliers and the delivery of tangible change closer to the entire NHS workforce.

From April 2005, the National Program joined forces with other Arms Length Bodies of the NHS to become 'Connecting for Health'. This effort released an estimated £500 million additional spending for the VLITP initiative. The new structure certainly became responsible for the delivery and support of the VLITP initiative as well as a number of the key services previously delivered by the NHSIA. It was hoped that the future concentration would be on integrating IT services at the local level to meet the demands of the frontline healthcare facilities.

An executive office for the NHS Information Authority (NHSIA) was formed in 2002, bringing together under the direct management of the Chief Executive of the authority the areas of corporate governance and clinical governance. The remit of the office was to interpret the NHSIA functions into business strategy and performance of IS for the NHS. As a result, the following few years were dedicated to organisational change for the directorate, which required the refocusing of some of its NHSIA objectives. Changes within the directorate structure involved the consolidation of the corporate communications' functions from other parts of the organisation and the creation of a corporate performance monitoring function to ensure the organisation remain on performance target set by the government. The NHS then had 180 thousand beds with 30 thousand administrators. The administration cost was 4.3 percent of the NHS budget.

To develop the standards or services required to meet the targets of the National Programme (an improvement to the ICRS project) and the wider health service, NHSIA has the ambition of building world-class applications, services and standards for the NHS. These are based on design authority, development of national services standards, development of local requirements, and development of IMandT capacity and/or capability and analysis service.

To meet these goals, the NHSIA set itself the following objectives:

- Develop core national services.
- Specify local application services.
- Support key service initiatives.
- Develop Health Informatics capability across the NHS.
- Develop a common framework for standards development activities.
- Develop required standards.
- Justify investments through business cases.
- Develop a strategic view of service requirements.

These could only be achieved with a consolidation of each departmental team's objectives. The system at the NHS requires that each departmental team continue to monitor its own objectives to ensure overall delivery of the consolidated plan. They are meant to achieve the above objective by the following means:

Figure 13.1. NHSIA communication concept summarizes its functions

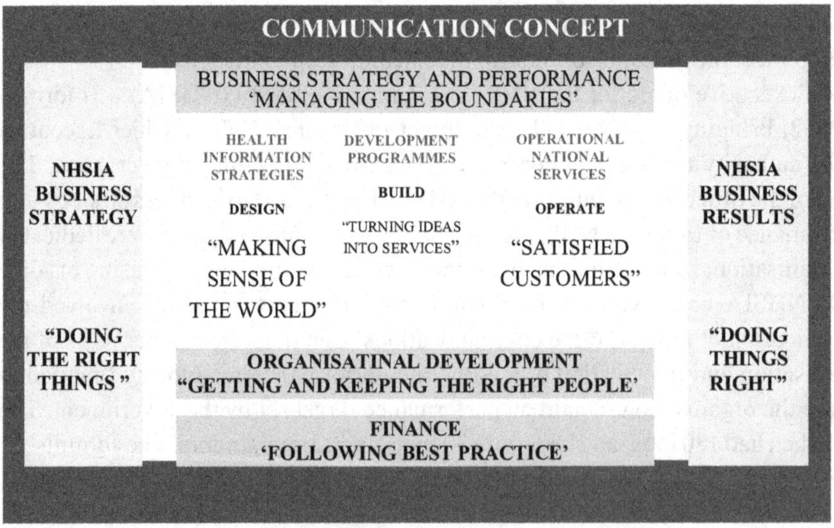

COMMUNICATION CONCEPT
BUSINESS STRATEGY AND PERFORMANCE
'MANAGING THE BOUNDARIES'
NHSIA BUSINESS STRATEGY
HEALTH INFORMATION STRATEGIES
DESIGN
"MAKING SENSE OF THE WORLD"
DEVELOPMENT PROGRAMMES
BUILD
"TURNING IDEAS INTO SERVICES"
OPERATIONAL NATIONAL SERVICES
OPERATE
"SATISFIED CUSTOMERS"
NHSIA BUSINESS RESULTS
"DOING THE RIGHT THINGS"
"DOING THINGS RIGHT"
ORGANISATINAL DEVELOPMENT
"GETTING AND KEEPING THE RIGHT PEOPLE'
FINANCE
'FOLLOWING BEST PRACTICE'

- Building and maintaining an understanding of the problems they are helping to resolve using consultation techniques involving the public and patients.
- Enabling the NHS to use information to improve the quality of patient care.
- Aligning their delivery in accordance with healthcare priorities, such as National Service Frameworks.
- Creating a reputation for delivering products and services that satisfy stakeholder requirements.

These interpret to an overall purpose of doing what could best be done nationally (see Figure 13.1). The mission of the NHSIA was to maximise the positive impact that information would make on improving the healthcare experience of the population. They also have the goal of being the national provider of information and infrastructure service to the NHS.

There have been several changes in the NHS strategic approach to IS over the last few years. This book uses a variety of empirical models to unpack the dynamics of institutional changes taking place in the NHS. Some of these changes include the longitudinal cradle-to-grave EHR and horizontal, local Electronic Patient Records (EPR) which provided the joint focus of the 1998 NHS IS strategy called 'Information for Health' (DoH, 2000). This strategy subsequently was referred to as integrated care record system (ICRS)—in a strategy document called 'Delivering 21st Century IT Support'. ICRS is a system of 'closely coupled' electronic care records at the heart of the NHS IS modernisation programme (Wanless, 2002).

In the 4th quarter of 2003, the initiative began to award contracts to private service providers under the National Programme for Information Technology (National Program). This is now the preferred nomenclature for the electronic care records that form the centrepiece of the NHS IS modernisation programme. Contracts were signed with Local Service Providers (LSP) and National Application Service Providers (NASP) to deliver the NHS care records service. The core of NHSIA strategy for National Program is to take greater control of the specification, procurement, resource management, performance management and delivery of the information and IS agenda. To deliver the necessary applications, services and IS infrastructure required within a timeframe that is both sensible and ensures value-for-money, National Program needs to run a procurement process that is as rapid as the NHS culture allows. Such a vibrant procurement process brought to the NHSIA the benefits of maintaining management focus and better engagement of prospective suppliers and the delivery of tangible change closer to the entire NHS workforce.

Using VLITP to Reform Very Large Organization

The research was conducted partly as a multilevel exploration of large IT implementation and has found differences in certain patterns as part of mutually reinforcing relationships with the organisational reform objectives. Moreover, these reform objectives were further reinforced by elements of larger institutional context. Upon conferring with contingency, configuration and congruence theories, a fit has been identified between multiple components of VLITP initiative: patterns of commitments, monolithic organisation and elements of the larger institutional context. The underlying idea behind these theories is that somehow the presence of one organisational structure enhances the impact of the others (Perlow et al, 2004).

Recognising that Nickerson and Zenger (2004) developed an alignment between governance forms and problems types, this thesis builds on the contention that these three governance forms differ fundamentally in their use of organisational features:

- Decision rights over the path of solution provided by IT solution.
- Communication channels to support knowledge transfer must be organisation specific.
- Incentives to motivate staff and customer when an organisation is restructuring need directional focus.

In this case, we have seen how certain organisational features are configured in unique and complementary ways to achieve individual organisational forms and functional prototypes. While cultural value orientations in the NHS alone cannot explain patterns of transforming an organisation form, they may still play a role in shaping the institutional context, which in turn appears to play a central role in shaping the re-structuring programme. Cultural value orientations should therefore be conceptualized as an additional outer ring in our theory of institutional structuration The theory emerging from the data in this paper also has important implications for better understanding and managing of externally led changes in multi-national organisation, government reform, as well as innovating large organisations. The theory here suggests that bringing and sustaining any of these patterns of interaction requires a great deal of understanding the patterns of interaction as well as the organisational, institutional and cultural context that enable and constrain the monolithic organisations.

Effective and Efficient Utilisation of Resources

VLITPs are increasingly becoming integral part of the infrastructure of an organization. Both successful and unsuccessful IT implementations can have an enormous impact and result in major operational and structural changes. VLITP management team cannot treat the implementation of hospital-wide systems in the same way as the purchase of a new MRI scanner or a departmental IT solution. The changes resulting from VLITP implementation quite often touch every member of staff in the host organization. They bring important trust-wide strategic issues to the fore. They also impact every political faction resulting to wide-ranging changes. VLITPs are therefore simply too important to delegate to junior business managers or IT departments in a healthcare institution. Likewise it is not good enough to have board members rubber-stamping every decision made by the VLITP management team. These are not just computer projects but organisational projects on a par with single site strategies and other major structural re-organisations.

VLITP is not as difficult as people think and one does not need to be a technical expert in order to direct it. The requirement is for excellent management skills and the ability to make things happen across a wide, varied and political organisation. Such skills are found mainly at board level, although the day to day running of a VLITP can be devolved. However, the project director, preferably the chief executive or finance director should aim to spend a considerable amount of time each week on the project and must be proactive rather than reactive.

VLITPs in the healthcare sector have proven to be very interesting, rewarding and shown to have massive corporate benefits, but they do not reach their full potential if the top medical practitioners do not get involved.

Human and Non-Human Resources

The NHSIA, like many VLITP management teams, operates within certain areas of constraint. Some of the significant types of constraints for the NHSIA in implementing the National Program project are:

- The skills, capabilities and readiness of "local" NHS institutions to implement "National" products and services provided. This problem can be demonstrated with the variances Local Area Networks and "Community Wide" Networks.
- The reliance on external, mostly commercial IT providers, to readily and speedily adopt the standards and data sets created for use without the availability of separate funds or suitable encouragement.

- The necessary limits and constraints upon funding leading to the inevitable decisions on the prioritisation of aspects of achieving "Building the Information Core: Implementing the NHS Plan."

Historically, the healthcare sector in the UK has been significantly under funded and has under performed in terms of IMandT investment. As a result of the process of annualized funding decisions, each year funding requests are made and decisions reached which might provide for growth or status quo or even reduction in IS budget. Due to this constraint, it is considered inappropriate for financial forecasts to be declared for the period covered by the National Program strategy. Given the variable and potentially volatile nature of the "market" in which the NHSIA operates, it has not considered significant alternatives to the National Program initiative. Preferring that its annual business plan will be flexible enough to accommodate changing circumstances, including funding limits and constraints and workforce plans.

However, the demised of several previous national initiatives may have been affected by the changing priorities of the NHSIA management, who were responding to further changes in the healthcare environment. The failure of this project may result from the IS not adapting to its environment technically, socially or economically. A responsive IS must be capable of rapid evolution in a changing environment.

Basic Economic Fundamentals

In Jessop's (1994) work looking at the nation-state and the rescaling of state activities due to complex processes of denationalisation, destabilisation and internationalisation, he considered sub-national scales as the breeding ground for regulatory experiments in the changed governance of economic development (Lipietz, 1994). Such assumptions were furthered emphasised by Jones (2001) and Ohmae (1995) that the nation-state has lost its role as the 'natural economic zone' and therefore no longer the central regulatory animation of economic development. Instead, it is assumed that economic development governance is increasingly conducted via sub-national, informal and formal networks as well as partnerships (Goodwin and Painter, 1996).

The broad governance approach does not offer analytical and explanatory help in this regard. Like the institutionalism literature, it gives prominence to the self-organising and relatively fluid nature of governance arrangements, including partnerships involved in public services and economic development (Jessop, 1997; Stoker, 2000). Governance approaches stress the role of change whilst downplays contingency and continuity in state apparatus and practices (Newman, 2001). Lowndes (2001) suggests the complex nature of change is reduced when pursuing this level of analysis. On the other hand, Newman (2001) argues that existing and emergent governance arrangements interact to produce tensions and disjuncture as differing norms, values and policy discourses come into play. Jessop also argues that governance ignores the distinctive constraints imposed by the self-organising dynamics and intersystem dominance of capitalism.

Studies by Scott (1996, 1998) have shown the importance of socio-institutional capacities that support and compose 'relational assets'. Not only for formal institutions but also informal institutions of norms, values and routines that underpin recurrent behaviour and learning, and which operate through inter-institutional relations (Storper, 1997). These works have shown that process of institutionalisation between agents, rather than their existence can create 'institutional thickness' supporting successful 'economies of association' (Amin and Thrift, 1994).

Partnerships form part of this complex governance environment (Bailey et al, 1995). Within the healthcare sector, they stem from a variety of processes, which have long been in operation in certain areas and have consistently formed part of British, European and American policies during the last two decades. The NHS has long been involved in partnerships as a mechanism in which to address particular issues, as part of general consultation and management strategies.

The NHSIA has a financial and procurement directorate that works on the basis of providing a service that is responsive to the customers and is of a consistently high quality. This is done through assisting its customers through the implementation of best procurement practice in accordance with the Authority vision and values. The provision of financial management and support is meant to help managers to make the right decisions. It also proves the NHS with the smooth processing of all financial transactions relating to IS.

NHSIA has a special health authority status, established with an overall remit to:

Improve patient care and achieve best value for money by working with NHS professionals, suppliers and academics and others to provide national products, services and standards, which support the sharing and most efficient and effective use of information (NHSIA, 2003).

This Case shows the need for integrating IS planning and strategic planning in the NHS, with attention placed on planning methods and their effects on patient care. Although the NHSIA wants to evaluate the success of planning especially after past projects have proven to fall short of at this level, there is a need to keep in mind the purpose of going through the planning process. The planning processes for the National Program seem to be very large, with extensive mapping and prioritisation of several key stages of achievements (NHSIA, 2003). This mapping can prove to be problematic for various reasons: user could be difficult to get involved, single processes could not be supportive of the NHS learning goals, and plans could even be too unrealistic.

One sees the NHS efforts for information inputs and planning resources that significantly related to the quality of the planning process for their VLITP. This case also shows that quality of integration mechanisms moderate the relationship between information input and the quality of the planning process. The quality of the planning process and the quality of implementation are significantly associated with the performance of a VLITP.

A distinction needs to be made between proprietary technologies and what might be called 'infrastructure technologies'. The former can be owned, actually or effectively, by a single Trust in the NHS—as long as they remain protected, proprietary technologies can be the foundations for long-term strategic advantages, enabling the Trust to reap higher benefits in supporting the healthcare process. The latter, in contrast, offer far more value when shared than when used in isolation. The characteristics and economics of infrastructure technologies make it inevitable that they will be broadly shared—that they will become part of the general health service infrastructure.

VLITP Management Challenges in Healthcare Sector

The initial challenge was the formulation of a national direction for IT that would meet the approval of a very bureaucratic Department of Health. Mr. Richard Granger, a former IT consultant, who has recently completed a small scale government IT project—The London Congestion Charging System—was appointed director for the newly dreamed national healthcare IT program and had instructed to work with a ministerial task-force to empower or deal directly with the strategic health authorities and manage those funds held centrally for this program. The newly appointed director also had to work with the Information Policy Unit and NHS Information Authority. There introduced the appointment of a chief information officer for each strategic health authority to ensure there were appropriate funding and effective IS management for every PCT and NHS Trust to implement and use the core IS solutions determined at national level. There had to be adequate capacity in PCT and NHS Trusts to implement the local elements of the VLITP. Mr. Granger divided the National Program project into regional work streams (see Figure 13.2) to develop the capability in the NHS to underpin the VLITP objectives.

Regional clusters (see Figure 13.2) were created after consultation with Strategic Health Authorities (SHAs) on how best to deliver local IT solutions as part of the VLITP. After much discussion, it was decided to split England into five geographic regions - each cluster comprising five, six or seven SHAs - who agreed to work together to take forward the procurement and implementation of the VLITP deliverables at local level. Applications delivered at a local level are the responsibility of five LSPs (Guah, 2008). The LSPs work closely with local NHS IT professionals and are overseen by a regional implementation director (RID) from the VLITP. The LSPs will ensure that existing local systems are compliant with national standards and that data are able to flow between local and national systems. To do this, the VLITP will deliver upgrades or replacements to hardware and software as appropriate and implement core local training for NHS staff.

All RIDs who led the input to negotiation and are leading the implementation process across their individual areas. An RID manages National Program support team and the relationship with the supplier, as well as co-ordinating deployment. An RID is part of the National Program team and reports to the National Programme implementation director, but is also responsible to the cluster board for delivery. The NHSIA had six key themes which describe its strategy for pursuing its Vision and Mission for this VLITP in the years ahead:

- **Stakeholders:** This theme focuses on engaging stakeholders appropriately and effectively so that the NHSIA can be certain that it is developing the right portfolio of products and services in response to the NHS needs. The VLITP management team emphasised a two-way communication with its stakeholders in the process of achieving the right balance in providing solutions that satisfy user requirements but also comply with Policy Direction and Corporate Governance.
- **National service delivery:** In the initial years, this theme was used to focus the VLITP on key services in support of modernizing the NHS so that its

Figure 13.2. Five english regional clusters with approximate population

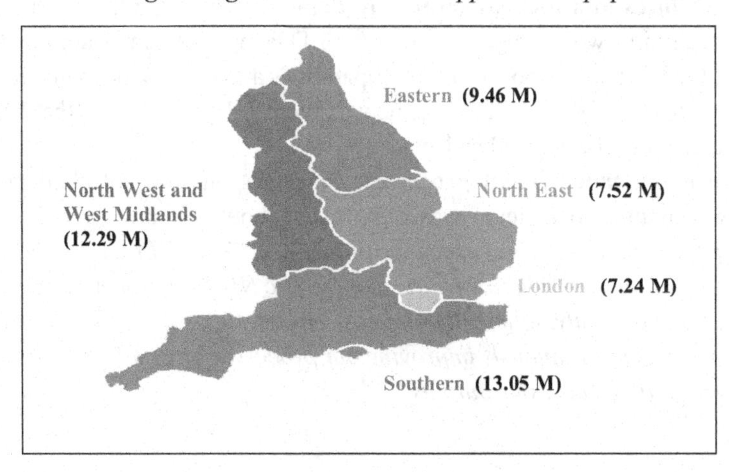

services are designed around the patient to improve equity of access and ensure a consistently high quality of healthcare services.

- **Information knowledge and management services:** The driver underpinning these services was to provide national information services that give NHS staff, patients and the public, access to relevant information and knowledge for decision-making. Within the first two years of the VLITP, the management team had already delivered the pilot of the national electronic library for health (NeLH), nhs.uk and National Analytical toolkits, amongst other products and services. In the following three years the management team needed to build on these successes and deliver the benchmarking and comparative analysis service; the national clinical audit service and the national knowledge service (NKS) for better health, amongst other things.
- **Health informatics:** This programme is directed at overcoming the lack of recognition for health informatics as a profession.
- **Electronic records:** The provision of electronic patient and health records (EPRs and EHRs) was at the very core of the NHS plan and information for health (IfH). To help expedite the programme a range of supporting products, guidance and templates were developed through the electronic records development and implementation programme (ERDIP).
- **Information infrastructure:** Work was underway, at an early stage, to put in place the basic infrastructure required to deliver a consistently high quality system across England and Wales. In subsequent years, the VLITP was expected to deliver an efficient, effective and fully supported integrated communications infrastructure for all those involved in the provision and receipt of health services. Elements of the deliverables included: managed services; generic tools; core service; access and security; physical networks and associated standards and polices.
- **Capabilities and infrastructure:** To deliver its strategic themes and their underpinning work programmes, the VLITP management team needed to have the right balance of skills and capabilities and the supporting infrastructure within the NHSIA. This theme was focussed on achieving that balance sometimes in the years ahead.
- **Improving management processes:** This was one area of challenge the VLITP management team was quite clear about. According to Professor Gwyn Thomas, then chief executive of NHSIA, *"The NHS is a Public Sector Organisation spending taxpayers' money. The NHSIA is committed to doing so as economically, efficiently and cost effectively as possible. We are also committed to continuously improving our processes by which we scope, plan, deliver and evaluate our outputs."*

- **Improving the working lives of the NHS staff:** It recognised that the NHS staff and their well-being were at the very heart of everything the VLITP management team was seeking to achieve in support of patients, clinicians and other healthcare practitioners. This theme focused on the initiatives it had in hand for ensuring that NHS staff continue to feel valued in return for their continued commitment to the organisational mission and goals.
- **Research and development within the NHSIA:** Alongside the commitment to deliver a balanced Portfolio through the pursuit of these 5 Strategic Themes and taking into account the types of constraints outlined above, the VLITP management team also committed to a level of applied and controlled research and development. This ongoing research and development will help to ensure that this and subsequent management teams can continue to contribute to the pursuit of improving patient welfare through the use of information management and technology

INSTITUTIONAL STRUCTURE OF THE NHS

Our look at institutionalism begins with a definition of institution, which is continuously being used to mean different things by researchers of political science, economics and sociology. These definitions go beyond that of Oxford dictionary (established, law, custom or practice). Lowndes (1996:182) presents it as "informal codes of behaviour, written contracts and complex organisations" with four elements:

- **A middle-level concept.** Institutions are devised by individuals and therefore constrain individuals' actions. Institutions here are seen as part of the broad social fabric and medium for individuals' day-to-day decisions and other activities. DiMiggio and Powell (1983) argue that institutions shape human actions, imposing constraints whilst providing opportunities for individuals.
- **Having formal and informal aspects.** Lowndes (1996) views institutions to involve formal rules or laws, which allows informal norms and customs to be practiced. That's because some institutions are not consciously designed nor neatly specified, yet part of habitual actions by its members. Such institutions may be expressed in organisational form and relate to the processes within.
- **Having legitimacy.** Legitimacy in institutions goes beyond the preferences of individual actors. Such preferences are valued in themselves and go beyond their immediate purpose and outputs.
- **Showing stability over time.** Lowndes (1996) views institutions as gaining their legitimacy due to their relative stability over time, and their links with a 'sense of place'.

New institutionalism theorists generally view institutions to have "the humanly devised constraints that shape human interaction" (North, 1990: 3) also refers to by March and Olsen, (1989:162) as "rules of the game" where organisations and individuals are constantly expected to play the game. Another stand taken by new institutionalism sees informal institutions (tradition, custom, culture and habit) are embedded in culture and conventions defined as behaviour structuring rules (March and Olsen, 1989; Mark and Scott, 1991; North, 1990; Peters, 1999). New institutionalism stresses embodied values and power relations of institutions together with interaction between individuals and institutions (Lowndes, 2002). Few authors have contested how far institutionalism concept can be expanded without becoming indistinguishable from other social facts and tautological (March and Olsen, 1998; Lowndes, 2002; Peters, 1999; Rothstein, 1996). Rothstein (1996:146) presents it as "focus in institutional analysis only on rules of behaviour that are actually agreed upon and followed by agents, whereby it is not relevant whether this agreement is reached explicitly or tacitly". Rothstein (1996) attempts here to distinguish between informal institutional rules and personal habits. Such distinction forms the basis for the definition of institution in this research where informal conventions and their impact upon the NHS and its partners are being explored. Our brief literature review uncovered extremely differing views on the new institutionalism theory. Peters (1999) identifies seven different versions: historical, rational choice, sociological, normative, empirical, institutions of interest representation and international institutionalism. Others like Mule (1999) economic and sociological and Thielemann (1999) argue for just two.

Table 13.3. Comparison in convergence of different institutional perspectives (adopted from Torfing & Sorensen, 2002)

	Historical	**Rational Choice**	**Sociological**
Analytical Aim	Explanation	Prediction	Understand
Focus	Meso-level	Micro-level	Macro-level
Institutions Affect	Balance of forces between organised interests	Range of options	Actors' identity, rationality and cognitive scripts
Institutional Change	Responses to external events	Intentional Reform	Institutional isomorphism
Institutional Formation	Result of codification of historic compromises	Voluntary contracts	Processes of diffusion and imitation
Roots	Marxist and Weberian political economy	Neo-classical economy	Phenomenology cognitive psychology post-structuralism
Theoretical Institution	Normative constructs of past inheritance	Rational constructs	Cultural constructs

Table 13.3 clarifies certain differences and overlaps in convergence variables by showing the contrasts between three major new institutionalism approaches. Torfing and Sorensen (2002) pointed out that due to all three approaches differing theoretical roots and development, it is quite difficult to agree on a clear definition and analysis of institutions. Such fundamental differences create a deep gulf between rational choice and sociological institutionalism—an ongoing unbridgeable debate. March and Olsen (1989) agree with the above tabulation because human behaviour is not rational but constantly influenced by the environment of the institution including, symbols, scripts and routines. These provide the filter for interpretation and affect the very identities and preferences of the actors.

Whilst taking into consideration the above ontological differences, they share the idea that institutions matter in the explanation of political life. It is much easier to find differences between the Old and New Institutionalism. Table 13.4 shows how new institutionalism has moved the analytical focus from formerly only organisations to rules. New institutionalists stress that institutions cannot be equated to political organisations (Peters, 1999). These are now regarded as set of rules, constraining the actors' behaviour. These rules work by determining appropriate behaviour—according to normative institutionalists (March and Olsen, 1989)—and provide inducements to behaviour in which individuals attempt to maximise their own utilities—according to rational choice institutionalist (Weingast, 1996). Institutions, here, provide the rules of the game (March and Olsen, 1989) and individuals—as well as organisations—are perceived as the players within that game. Compared to the old, new institutionalism expands the definition of what comprises an institution to include informal rules, conventions, norms and values. Everything that guides individual behaviour becomes subject of new institutional research (Peters, 1999).

New institutionalism has also shifted the focus of analysis to a value-critical stance. They are interested in exploring how institutions embody and structure

Table 13.4. Comparison of old and new institutionalism (Lowndes, 2002)

	OLD	NEW
Research Focus	Organisations	Rules
Definition	Formal institutions	Rules are expanded to informal and conventions
Characterisation	Static Processes are facts of life	Institutions have dynamic processes.
Roles of Value	Submerged	
Understanding	Holistic	Differentiated
Position in Space and Time	Free Standing	Embedded

societal values. Though the old institutionalist pursued holistic analysis of whole government systems, the new version focuses upon individual institutions of political life. New institutionalists are interested in exploring how institutions embody and structure societal value unlike focusing on a particular set of values and model of government. They explore how institutions are embedded.

Institutional ideas and variables are frequently incorporated into research designs as either supplementary or alternative explanations to other theoretical perspectives. The discussion below describes compelling evidence that institutional theory broadens our understanding of how organizations interact with their environments. The most common theory combinations are with population ecology, resource dependence, contingency, and strategic choice theories. These theory combinations help to explain the diffusion of innovation, organizational death rates, performance, the "liability of newness," board involvement, organizational structure, isomorphism, and inter-organisational relations.

One area where combining institutional theory with other theoretical explanations of healthcare IS strategy could enhance understanding is the diffusion of innovation. Tolbert and Zucker's (1983) study regarding the adoption of civil service reform identified a difference between the motives of early and late adopters. Early adopters "rationally" select an innovation to increase organizational "performance." On the other hand, later adopters pursue an innovation to gain legitimacy. This pattern appears conceptually in the explanation of managerial technology by both Abrahamson (1991) and Currie (2004). Both papers incorporate the ideas of coercive and mimetic isomorphism and the pursuit of legitimacy as potential explanations for the adoption of innovative technologies.

VLITPs are being implemented in the healthcare sector because it is one of the most lucrative sectors for deploying, hosting and integrating web-enabled software applications. This case illustrates some of the key issues, listed below:

- First, as a national healthcare provider, IS strategy in the NHS is complex and multi-dimensional. This is partly due to the NHS being a monolithic organization with multiple decision-makers across managerial and clinical levels and units.
- Second, institutional approach deals with the choices made in response to or in compliance with the organization's institutional environment. The chapter denotes whether this VLITP was able to enhance IS efficiency and operations in the host organization.
- Third, whether this case provides evidence from the data collected during the study and deduce different scenarios affecting the implementation of VLITPs in healthcare institutions.

The broad approach of new institutionalism claims that institutions constrain behaviour (DiMaggio and Powell, 1983; Peters, 1999; Scott et al, 2000). By expressing institutions as embodying values and power structure rather than new neutral, a new research field in social science has been unveiled. In addition to exploring the affect of institutions on individual behaviour, new institutionalists explore how such interaction takes place between the institutions and individuals. New institutionalism is also expressed from two different points to view: rational choice and normative. Such bipolar nature allows scholars to investigate human behaviour in institutional analysis from two different positions. The rational choice angle regards preferences as exogenous and the normative one views them as endogenous. March and Olsen (1989) stress the former explain actor behaviour by following a 'logic of consequentiality' and the latter by means of a 'logic of appropriateness'. The institution is seen to provide the 'rules of the game', the structure and behaviour. These rules are interpreted in two ontologically different ways. Normative institutionalist see an individual's action as the matching of a situation to the demands of a position, and therefore obligatory. In being 'in touch with his identity' an actor fulfils the obligations of a role in a situation and follows intentionally and institutionally defined 'logic of appropriateness'. Rather than contemplating the consequences, he does what he considers to be the most appropriate action (March and Olsen, 1989).

This series of thoughts contrasts with anticipatory action associated with 'logic of consequentiality'. Matching of subjective preferences with expectations about consequences drives the actor's behaviour. The individual, who is 'in touch with reality', lists its alternatives for action, and decides which alternatives would best maximise utility and then acts accordingly. But for rational choice institutionalist, individual's behaviour and action is wilful and anticipatory and thus, in choosing the best consequences he follows 'logic of consequentiality (March and Olsen, 1989). The two contrary behavioural assumptions do not imply using different approaches in institutional analysis. Despite ontological differences, they both have the idea in common that action and behaviour is rule-bound. The challenges for new institutionalist researcher lies in identifying the rules that are responsible for a certain behaviour or action (Kjaer and Pederson, 2001).

Decision Making Environment

At a time in the NHS when there are very real and pressing problems those UK healthcare faces, it may appear to be the wrong time to claim that we need good information systems. But, in fact we do. The primary care institutions are coming under increasing pressure to meet government standards as policies are being implemented to facilitate patients moving around freely to any GP. If specific secondary care facilities are identified as failing repeatedly, patients will soon have the choice in moving to different hospital. Such failing NHS facility will have real incentive to change. The question is, "Change how?" But even with the best intentions of wanting to improve healthcare, attempts to change will be trial and error if there is no alternative to predict the consequences of such changes. The alternative to trial and error is valid healthcare IS. The NHS needs to go through a process of learning that emerging technologies do change the practice for running a public service organisation. The NHSIA has been willing to cannibalise existing services and engage in strategic alliances with private service providers in the National Program initiative. The process involves the commitment of significant resources for various alliances to explore a wide range of opportunities and leverage those that succeed. The NHSIA intension is to speed implementation by developing incentives to attract and retain service providers of proven track records in health care IT.

Organising Vision

The thesis elaborates Swanson and Ramiller (1997) use of the term *organising vision* to characterise a variety of IT innovations being developed within the NHS to improve the quality of health care and services provided to the British public, reduce the cost of provision and assist the medical practitioners in gaining easier and faster access to patient records. Currie (2004) re-defines an organising vision that incorporates not only information technologies, but also assumptions about organising business practices—healthcare delivery in the case of the NHS—and institutions to take advantage of IT capabilities. The organising vision for National Program has been interpreted by our interviewees to incorporate ideas about coordinated clinical care, reduced medical errors, and improved compliance to NICE standards and DOH guidelines. Such organising vision also facilitates the Trusts managements' own interpretations of the social and organisational implications for National Program, which can be referred to as innovation, legitimisation of IT diffusion and adoption at local level, and mobilisation of resources in support of the e-government initiative. What has clearly come out of the research results is that organising visions for National Program is the stimulation of interest and investment in NHS IT, despite uncertainties about the total costs, anticipated benefits and long-term implications for future levels of GP internal communications to the patients.

Finding suitable IS for the NHS is complex with regards to the organisational technologies involved. Their applications, uses, limitations and implications are not clear-cut nor could benefits from use ever be assured. Clear and convincing ideas about how to use IS in the NHS to improve quality, reduce costs and delivery access to patients' records are critical to promoting adoption and guiding successful implementation of National Program. Playing a pivotal role in all these are important stakeholders that includes medical practitioners, leaders of different sub-organisations in the NHS, the DoH and other regulatory agencies, as well as IT service providers and consultants in the private sector. Their level of participation will determine the shape and outcomes of any possible innovations emerging from National Program. Understanding the social construction and interpretive processes through which the National Program has emerged, developed, and been communicated, is key to achieving anticipated outcomes of adoption and informing healthcare reform policies.

According to Currie and Willcocks (1998), IS role is changing from one of automating support functions in a quest for greater efficiency to one where core business processes are being transformed and wider strategic benefits are sought. Hammer (1990) considers IS as the key enabler of organisation process, which he considers as "radical change". He prescribes the use of IS to challenge the assumptions inherent in an organisation processes and procedures that have existed since long before the advent of modern computer and communications technology. According to Swanson and Ramiller (1997: 458), an organising vision serves several distinct functions in the adoption and diffusion process of such innovation process:

- The vision presents the host organisation an ongoing *interpretation* of IT innovation which takes into account what the vision is about and how the vision could be used. The vision of National Program helps answer the question of what would be different in the NHS if the innovation is adopted. The organising vision provides the rationale for adopting and using the emerging innovation, in terms meaningful to decision makers and general management within the NHS.
- The case shows by doing the above, the vision will *legitimise* the emerging innovation and fuel its further adoption.
- This organising vision seems to have *mobilised* the private sector in ways that is facilitating development and diffusion of other major IT project in UK public sector. As a result of the formation of consortiums, standards have been indirectly developed for the facilitation and development of complementary products and reassuring potential adopters in the current government—as a major customer.

Within this discourse, the organising vision takes shape as a solution to general problems or issues which have been demonstrated in the Case. Further problems may emerged for which an organising vision provides a solution by refining, extending, and sometimes redirecting as more local Trusts gain experience with the innovation, and these experiences filter into the discourse.

The result of this research confirms Swanson and Ramiller (1997: 462) suggestion that to effectively promote adoption and diffusion of an IT innovation, an organising vision must be distinctive and plausible. National Program has given the message that a relatively new and massive IT implementation project is taking place which is going to affect the institutional operations of the NHS. That project—being a major part of the NHS modernisation, reform, re-computerisation, etc—could conceivably solve most of the problems associated healthcare delivery. One caution here is that an organising vision that falls very "far out" may fail to enrol supporters, as will visions that are too mundane. Nevertheless, the National Program as an organising vision is playing an important role in the early stages of diffusion of NHS reform—partially because the understanding of the innovation and its value are vague.

The Case attempts to answer major questions relating to certain key areas for analysis. What are the key issues ("the business problems") that the organising vision is identified with? What are the core information technologies in the vision? What are the implicit and explicit assumptions about how the NHS and healthcare sector will be structured to apply IT and address the business problems? As the various sub-organisations within the NHS gain and share their experiences with this innovation, these elements may change dramatically. It points out that organising vision often build on earlier visions or result from the subsuming of other visions. National Program build on earlier IT innovations in the NHS (including Electronic Health Record project, National Care Record Service, Electronic Transmission of Prescriptions, Picture Archiving and Communications Systems, ICRS, etc) to provide a new computer systems and services that will connect over 100,000 doctors, 380,000 nurses and 50,000 other health professionals. Examining how the National Program vision relates to and also differs from, other visions will highlight the dynamic influences of the *business problems*, technological capabilities, and medical practitioners' interpretation of and experiences with the National Program vision.

Assumptions about IT and the NHS are interrelated in the organising vision for National Program in complex ways. Most discussions of the benefits of National Program are founded on these few common assumptions:

- The collection and sharing of patients' records electronically
- Reduction of medical errors
- Avoidance of duplication of services due to lack of information
- Reduction of misinformation and misinterpretation of patient records
- Fast and easier access to patient records at the point of healthcare delivery

One major assumption is this VLITP was to introduce a more informed and cost-effective decisions within the NHS healthcare processes. One justification for the VLITP has included the idea of sharing critical healthcare data with patients and encouraging the active involvement of patients in the management of their own health. This is primarily due to the advent of many healthcare websites on the Internet.

The organising vision theory emphasised in Currie (2004) highlights the functionality. A VLITP vision must provide (interpretation, legitimization, mobilization) and the characteristics that the vision must have to promote adoption (interpretability/informative, plausibility, importance, distinction). The case has brought forth the discourse related to VLITP that provides interpretations of problems in the healthcare process and identifies opportunities to use IT for national database of patients' records. Issues such as lack of standards for interoperability and misaligned incentives for adoption have been identified and policy recommendations formulated. In these ways the organising vision for VLITP becomes more interpretable. The case shows evidence of legitimization in the form of positive experiential reports, and cost-benefit analysis. Mobilization was evident in the formation of regional Trusts collaboration to develop regional IT training and other interest groups, to prepare policy recommendations for the VLITP management team and for service providers, and to address standards issues with NHS IT.

The necessary programme architecture to achieve the VLITP objectives to benefit the NHS is illustrated in Figure 13.3. The key technical issue faced by the VLITP management team was deciding which object technology to use for the National Program. There are two competing standards, DCOM and CORBA, support by Microsoft and their commercial competitors respectively. This is where Java comes into the picture. It is an object oriented programming language with a number of built in features that enable Internet application development and deployment.

The weaknesses of these technologies lie in their immaturity (Hagel III, 2002). Staff with the relevant skills are hard to find and expensive. The technology is still

Figure 13.3. National strategic architecture: panorama of NHS IS

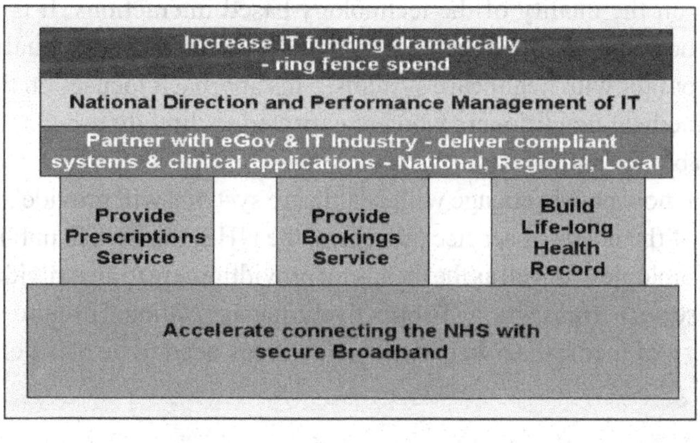

evolving so there is a risk that the NHS could face certain drawbacks, thereby affecting the integrity of data involved. For corporate use, there is still considerable suspicion about the safety of downloading software via the Internet. Any use of downloaded Java or Active X is still blocked by many Trusts. For patients, the lack of bandwidth on the public Internet will limit take up. A set of Java beans or Active X components can easily exceed 2MB in size: in today's environment, that could take 30 minutes or more for a user to download.

Given web services technology, upgrades of applications can happen immediately so it no longer takes 1 month to roll out 100 PCs in a single Trust. New users (i.e. temporary staff) can immediately be given access to the applications they need. Email is even quicker and in many cases more convenient than contacting people by phone. Sending documents, drafts of contracts, patients can request information and receive a reply to be acted upon when convenient. Patients can immediately access the information on scheduling and drugs availability, whilst information can updated on a daily or weekly basis. It becomes easier to reduce inventory and to manage those costs if a Trust has a more up to date picture of demand.

QUALITY OF HEALTHCARE DATA

Quality is a fundamental challenge in an ever-changing society, particularly with regards to IS which is used in all areas of social and professional lives. LeRouge et al (2004) indicate the imperatives of understanding IS quality to be determining how one can best manage service encounters for an organisation like the NHS to produce desired outcomes. LeRouge and her colleagues also emphasised that service encounters are critical interactions between service providers and recipients because it demonstrates an organisation's capability to fulfil its mission and shape consumers' impressions of the organization. The effectiveness of National Program interaction depends not only on the use of the resulting state-of-the-art technology but also on the quality of the technology-based interactions. It is necessary to consider the socio-technical approach to understanding system quality during patients' encounter with healthcare systems. This approach focuses on the perceptions of the medical practitioners who are expected to directly use the technology and accountable for patient care.

Looking at how people engage with healthcare systems will provide an interesting instance of the nexus of service providers, the NHS, the British public, and the resulting technologies; as well as the means for providing care to an individual during a health care/service 'transaction'. To effectively manage National Program influence on the delivery of medical service, two quality areas need to be clarified:

1. **Quality in functional:** This addresses personnel, technology, physical environment and patient acting as quality-generating resources during the encountering process. The functional view of quality takes into consideration the unit of software including graphical user interface, the hardware, embedded systems for control and regulation of peripherally technical processes. It also involves communication with other IS, and associated social action system of persons, who are acting with the technology and other people.
2. **Quality in desired clinical encounter:** This results to diagnostic accuracy, diagnostic impact, and therapeutic impact as well as other contributing success factors including patient and direct medical practitioner satisfaction. The recognition of technology-based service encounters as complex engagements of a socio-technical system is necessary to make significant progress in addressing challenges regarding success in both research and practice of system encountering.

Quality within the VLITP context requires the recognition that quality cannot be expressed in a singular vernacular and no perspective alone provides a complete definition. The patient and medical practitioners serve as key participants throughout the encounter process, though they may differ in perspectives of the system's quality. The VLITP management team perspective of the quality of project was to elucidate insight from the central figure of responsibility and encountering activity. The patient perspective came from a benefit angle. The collective exploration of multiple viewpoints was critical to success if high-quality health service was to be the ultimate goal of the VLITP.

For medical practitioners, explicit representation of the quality attributes of National Program from the perspectives of both patients and the NHSIA provides insights essential to implementation, utilisation, and common understanding. Without an understanding of system quality in the VLITP, the potential for successful implementation and utilisation, as well as knowledge building will be diminished. A leading service provider to most primary care trusts was refusing to collaborate with LSPs and therefore didn't participate in the VLITP project within the first eighteen months without satisfactory answers to the following questions:

- What quality attributes will the VLITP contribute, from the perspectives of both service providers and patients?
- Are there differences, of relative importance, among attributes from the anticipated technologies and EMIS' existing plans for the NHS?

The ICRS will be a broad, continuously expanding and maturing portfolio of systems and services to create, store, share, transfer and give access to health re-

Figure 13.4. PSP in support of the patient's healthcare process

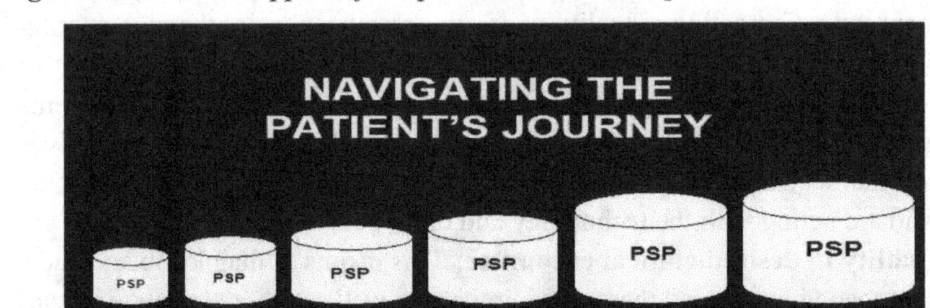

cords. A key part of this consists of the tools to support the patient journey along care pathways (see Figure 13.4). It will include each organisation's patient records and a nationally shared summary of patient information, called the 'NHS spine'.

CHAPTER SUMMARY

The NHS is clearly a highly complex organization and the theory being proposed in this thesis suggests an understanding of systems complexity in the NHS. Due to the multiplicity of systems in the NHS, there are various internal organizations (i.e. Cancer Society, National Institute for Clinical Excellence, etc.) requiring technology to facilitate meaning at all times. The potential for current service providers to satisfactorily provide the required service to the NHS does exist; though certain concentrated improvements need to be made to their services. The critical nature of NHS services requires the service providers to provide a higher value and managed service.

This VLITP has strived to accelerate the diffusion of healthcare knowledge in the UK. Medical knowledge is rapidly changing from breakthroughs that accelerate the introduction of new medications. However, even well synthesised knowledge faces many hurdles to being used in clinical practice. Balas et al (2000) estimate it to take 17 years, on average, for evidence to be integrated into clinical practice. That's mainly due to the enormous amount of information available to healthcare organisations; healthcare professionals find it increasingly difficult to keep current with new findings in their clinical practices.

The NHS has long been expected to reform and improve it processes. This chapter suggests VLITP as a catalyst to de-institutionise and change very large institutions. While the adoption of web-enabled solutions using VLITP presented the NHS with an opportunity to benefit from certain emerging technologies over a period of time, the data support another very interesting conclusion. The Department of Health's top-down approach designed to be compatible with professional interests, is unseating professional dominance indirectly through the profusion of conflicting governance structures that followed the National Program initiative. This situation is being legitimated by the rise of medical costs. The business interests of IT service providers are also the apparent beneficiaries rather than the cause of the decline in professional dominance.

It has never been easy to implement change in the NHS. However, early adoption success of the applications resulting from this VLITP shows that the NHS is unlikely to adopt mission-critical solutions until issues of data security and reliability are overcome (Guah and Currie, 2004). At the present time, the inhibitors of the National Program model for the NHS are twofold:

- First, the NHS is investing in new initiatives to improve efficiency and effectiveness. Software vendors, offering National Program solutions to the NHS, need to understand the decision-making hierarchies and levels within the NHS and within government before targeting their solutions to specific business or vertical areas. The software applications resulting from the National Program project will need to be evaluated through measurable performance indicators. This is because a web-enabled delivery system does not in itself add any additional business or operational advantage to the NHS. Over time, the NHS may need to pursue National Program aggregation strategies, which fully integrate disparate software applications. However, this will require an NHS IS strategy, which fully captures the complexity of using multiple vendors.
- Second, emerging technologies have the potential to remake entire industries and render established strategies obsolete. For large organisations, like the NHS, however, emerging technologies often have a traumatic impact. Such organisations often feel they must utilize any new technology that emerges. Their first reason is defensive, driven by the belief that newcomers are plotting to use the new functionalities to attack their core business. Their second reason is the converse of the first: if the emerging technology realizes its potential, it will be too attractive to ignore. In the thesis, the researcher addresses the questions of why the NHS has so much difficulty with emerging technologies.

This chapter has explained the NHS drive to invest in the implementation of a VLITP to improve IS support for its core healthcare processes. Despite much IS investment, the NHS has been slow to adopt new IS-enabled methods and practices. This has been echoed at local Trusts level as well as the NHSIA, as organisation-wide reform has been met with an alleged unwillingness by many parts of the NHS to embrace change. Inefficient and outdated practices still permeate the NHS, as many IS vendors find it increasingly difficult to penetrate the various decision-making hierarchies (Wanless, 2002).

It is clear from the discussions in this chapter that no one generic strategy for VLITP has been successful in the NHS. In its strategy, the Connecting for Health has considered patient care objectives with a focus on healthcare processes. The chapter has also shown that attempts to add on an e-commerce facet to the NHS traditional business structure without re-engineering their traditional processes are not likely to succeed. Moreover, since NHS systems tend to be pervasive and are continuously evolving, an integrated approach is called for.

REFERENCES

Armenakis, A. A., & Bedeian, A. G. (1999). Organizational Change: A Review of Theory and Research in the 1990s. *Journal of Management, 25*(3), 293-315.

Atkinson, C. J., & Peel, V. J. (1998). Transforming a Hospital through Growing, not Building, an Electronic patient Record System. *Methods of Information in Medicine*, *37*(3), 285-293.

Audit Commission. (1995). *For your information, a study of management and systems in the acute hospital.* London, UK: HMSO.

Barley, S. R., & Tolbert, P.S.(1997). Institutionalization and Structuration: Studying the Links between Action and Institution. *Organization Studies*, *8*(1), 93-117.

Bowersox, D. J., Closs, D. J., & Stank, T. P. (2000). Ten mega-trends that will revolutionize supply chain logistics. *Journal of Business Logistics*, *21*(2), 1-16.

Brown, S. (2001). NHS finance: The issue explained. *The Guardian. May 30*, 2.

Byrd, T., Sambamurthy, V., & Zmud, R. (1995). An examination of IT planning in a large, diversified public firm. *Decision Sciences*, *26*(1), 49–74.

Cavaye, A., & Christiansen, J. K. (1996). Understanding IS Implementation by Estimating Power of Subunits. *European Journal of Information Systems, 5*(4), 222-232.

Chen, Y., & Perry, J. L. (2006) Managing Government and Healthcare IT Outsourcing in Europe, EU Directive 2002/58/EC, accessed March 20, 2005, available at www.cdt.org.

Connecting for Health (2004). Business Plan. Accessed June 10, 2005, available at http://www.connectingforhealth.nhs.uk.

Currie, G., & Suhomlinova, O. (2006). The impact of institutional forces upon knowledge sharing in the UK NHS: The triumph of professional power and the inconsistency of policy. *Public Administration, 84*(1), 1-30.

Currie, W. L., & Guah, M. W. (2006) IT-Enabled Healthcare Delivery: The UK National Health Service. *Information Systems Management, 23*(22), 7-22.

Davern, M. J., & Kauffman, R. J. (2000). Discovering Potential and Realizing Value from Information Technology Investments. *Journal of Management Information Systems, 16*(4), 121-143.

Department of Health. (2002). Delivering 21st century IT support for the NHS. Department of Health, London, UK: HMSO.

Dewire, D. T. (2000). Application Service Providers. *Information Systems Management, 17*(4), 14-19.

Doherty, N. F., King, M., & Marples, C. G. (2000). The impact of Hospital Information Support Systems on the Operation and Performance of Hospitals. *Information Systems Review, 1*(1), 97-107.

Douglas, T. J., & Fredendall, L. D. (2004). Evaluating the Deming Management Model of Total Quality in Services. *Decision Sciences, 35*(3), 393-422.

Eisenhardt, K. M. (1989). Building Theories form Case Study Research, *Academy of Management Review, 14*(4), 532-550.

Fredrickson, J. (1984). The comprehensiveness of strategic decision processes: Extension, observations, future directions. *Academy of Management Journal*, 27(3), 445–466.

Galpin, T.(1996). *The Human Side of Change: A Practical Guide to Organization Redesign.* San Francisco: Jossey-Bass.

Guah, M. W., & Currie, W. L. (2006). *Internet Strategy: The Road to Web Services.* Hershey, PA: IRM Press.

Guah, M. W., & Currie, W. L. (2007). A National Program for IT in the Organisational Field of Healthcare: An Example of Conflicting Institutional Logics. *Journal of Information Technology*, Forthcoming.

Gattiker, T. F., & Goodhue D. L. (2005). What happens after ERP implementation: understanding the impact of inter-dependence and differentiation on plant-level outcomes. *MIS Quarterly, 29*(3), 559-585.

Hagel, J. III (2002). *Out of the Box: Strategies for Achieving Profits Today and Growth Tomorrow through Web Services.* Boston, MA: Harvard Business School Press.

Hagel, J. III., & Brown, J. S. (2001). Your next IT Strategy. *Harvard Business Review, 79*(9), 105-113.

Hakonsson, D. D. (2006). How misfits between managerial cognitive orientations and situational uncertainty affect organizational performance. *Simulation Modelling Practice and Theory, 14*(1), 385-406.

Handfield, R. B. (1994). U. S. Global Sourcing: Patterns of Development. *International Journal of Operations and Production Management, 14*(6), 40-51.

Hao, Q., Shen, W., & Wang, L. (2006). Collaborative manufacturing resource scheduling using Agent-based Web Services. *International Journal of Manufacturing Technology and Management, 9*(3), 309-327.

Heathfield, H., Pitty, D., & Hanka, R. (1998). Evaluating information technology in healthcare: barriers and challenges. *British Medical Journal, 316*, 1959-61.

Hitt, L. M., Wu, D. J., & Zhou, X. (2002). Investment in Enterprise Resource Planning: Business Impact and Productivity Measures. *Journal of Management Information Systems, 19*(1), 71-98.

Institute of Medicine (2002). *Crossing the quality chasm: A new health system for the 21st century*, Committee on Quality Health Care in America, Washington, DC: National Academy Press.

Jacobs, F. R., & Bendoly, E. (2003). Enterprise Resource Planning: Developments and Directions for Operations Management Research. *European Journal of Operational Research, 146*(2), 233-240.

Jiang, J. J., & Klein, G. (1999). Risks to different aspects of systems success. *Information and Management, 35*(10), 263-272

Klecun-Dabrowska, Ela, & Cornford, T. (2000). Telehealth acquires meanings: Information and communication technologies within health policy. *Information Systems Journal, 10*(1), 41-63.

Koufteros, X., Vonderembse, M., & Jayaram, J. (2005). Internal and external integration for product development: The contingency effects of uncertainty, equivocality and platform strategy. *Decision Sciences*, *36*(1), 97-134.

Kumpers, S., Van Raak, A., Hardy, B., & Mur, I. (2002). The influence of institutions and culture on health policies: Different approaches to integrated care in England and The Netherlands. *Public Administration*, 8(2), 339-358.

Lacity, M. C., & Willcocks, L. P. (2006). Transforming back offices through outsourcing: Approaches and lessons. In Leslie P. Willcocks and Mary C. Lacity *Global Sourcing of Business and IT Services*, New York, NY: Palgrave Macmillan, 97-113.

Lieb, R. C. (1992). The Use of Third-Party Logistics Services by Large American Manufacturers. *Journal of Business Logistics*, *13*(2), 29-42.

Lowson, R. (2002). 'Assessing the Operational Cost of Offshore Sourcing Strategies.' *International Journal of Logistics Management, 13*(2), 79-89.

Lucas, H. C., Jr., Walton, E. J., & Ginzberg, M. J. (1988). Implementing Packaged Software. *MIS Quarterly, 2*(4), 537-549.

Majeed, A. (2003). Ten ways to improve information technology in the NHS. *British Medical Journal*, *326*, 202-206.

Maltz, A. B., & Ellram, L. M. (1997). Total cost of relationship: an analytical framework for the logistics outsourcing decisions. *Journal of Business Logistics*, *18*(1), 45-66.

Mark, A., Pencheon, D.. & Elliott, R. (2000). Demanding Healthcare. *International Journal of Health Planning and Management*, *15*(1), 237-253.

Markus, M .L., & Tains, C. (2000). The Enterprise System Experience- From Adoption to Success," In *Framing the domains of IT management: projecting the future through the past*, R. W. Zmud (eds.), Cincinnati: Pinnaflex, 173-207.

McNulty, T., & Ferlie, E. (2004) Process transformation: Limitations to Radical Organisational Change within Public Service. *Organization Studies*, *25*(8), 1389-1412.

Mitchell, V. L., & Zmud, R. W. (2006). Endogenous Adaptation: The effects of Technology Position and Planning Mode on IT-Enabled Change. *Decision Sciences*, *37*(3), 325-355.

Nidumolu, S. (1995). The effect of coordination and uncertainty on software project performance: Residual performance risk as an intervening variable. *Information Systems Research*, *6*(3), 191–219.

Orlikowski, Wanda J., & Tyre, M. J. (1994). Windows of Opportunity: Temporal Patterns of Technological Adaptation in Organizations. *Organization Science*, *5*(1), 98-118.

Petersen, K. J., Frayer, D. J., & Scannell, T. V. (2000). An Empirical Investigation of Global Sourcing Strategy Effectiveness. *Journal of Supply Chain Management 36*(2), 29-38.

Pettigrew, A. (1990). Longitudinal Research on Change: Theory and Practice, *Organization Science, 1*(3), 267-292.

Pettigrew, A. M. (1987). Context and Action in the Transformation of the Firm. *Journal of Management Studies, 24*(6), 649-670.

Pich, M., Loch, C., & De Meyer, A. (2002). On uncertainty, ambiguity, and complexity in project management. *Management Science*, *48*(8), 1008–1023.

Robey, D., Ross, J. W., & Boudreau, M. (2002). Learning to Implement Enterprise Systems: An Exploratory Study of the Dialectics of Change. *Journal of Management Information Systems*, *19*(1), 17-46.

Reyes, P. M. (2006). A game theory approach for solving the transshipment problem: a supply chain management strategy teaching tool. *Supply Chain Management: An International Journal*, *11*(4), 288-293.

Sabherwal, R., & Chan, Y. (2001). Alignment between business and IS strategies: A study of prospectors, analyzers, and defenders. *Information Systems Research*, *12*(1), 11–33.

Sabherwal, R., & King, W. (1992). Decision processes for developing strategic applications of information systems: A contingency approach. *Decision Sciences*, *23*(4), 917–944.

Scott, W. R., Ruef, M., Mendel, P. J., & Caronna, C. A. (2000). Institutional Change and Healthcare Organizations: From Professional Dominance to Managed Care. Chicago, IL: University of Chicago Press.

Sink, H. L., & Langley, C. J. (1997). A managerial framework for the acquisition of third-party logistics services. *Journal of Business Logistics, 18*(2), 163-189.

Soh, C., Kien, S. S. and Tay-Yap, J. (2000). Cultural Fits and Misfits: Is ERP a Universal Solution? *Communications of the ACM*, *43*(4), 47-51.

Spencer, B. A. (1994). Models of Organization and Total Quality management: A comparison and critical evaluation. *Academy of Management Review, 19*(3), 446-447.

Stratman, J. K., & Roth, A.V. (2002). Enterprise Resource Planning (ERP) Competence Constructs: Two Stage Multi-Item Scale Development and Validation. *Decision Sciences, 33*(4), 601-628.

Sussman, S. W., & Siegal, W. S. (2003). Information influence in organizations: An integrated approach to knowledge adoption. *Information Systems Research, 14*(1), 47–65.

Udo, G. G. (2000). Using analytic hierarchy process to analyze the information technology outsourcing decision. *Industrial Management and Data Systems, 100*(9), 421-429.

Venkatraman, N. V. (2004). Offshoring without Guilt. *Sloan Management Review, 45*(3): 14-16.

Van de Ven, A. H., & Huber, G. P. (1990). Longitudinal Field Research Methods for Studying Processes of Organizational Change. *Organization Science* 1(3), 213-219.

Van Hoek, R. (2001). E-supply chain—virtually non-existing. *Supply Chain Management: An International Journal*, 6(1), 21-28.

Wanless, Derick (2002). Securing our Future Health: Taking A Long-Term View. Final Report of an Independent Review of the long-term resource requirement for the NHS. London, UK: HMSO.

Whitley, R. (2003). The institutional structuring of organizational capabilities: the role of authority sharing and organizational careers. *Organization Studies,* 24(5), 667-696.

Wei, H-L E., Wang, T. G., & Ju, P-H (2005). Understanding misalignment and cascading change of ERP implementation: a stage view of process analysis. *European Journal of Information Systems, 14*(4) 324-334.

Wenger, E. C., & Snyder, W. M. (2000). Communities of Practice: The Organizational Frontier. *Harvard Business Review,* 78(1), 139-147.

Willcocks, L., & Choi, C. J. (1995). Co-operative partnership and "total" IT outsourcing: From contractual obligation to strategic alliance? *European Management Journal*, 13(1),67-78.

Willcocks, L., & Currie, W. L. (1997) Pursuing the re-engineering agenda in public administration. *Public Administration*, 75(4), 617-650.

Wilson, E. V., & Lankton, N. K. (2004). Interdisciplinary Research and Publication Opportunities in Information Systems and Healthcare. *Communications of the Association for Information Systems*, 14, 332-343.

Chapter XIV
Case Study II: RFID—A Technology for Enterprise Systems in the Airlines Industry

ABSTRACT

VLITP can shift the direction of organizations by introducing new systems and emerging technologies that can serve as a trigger for change to the entire business strategy of an organization. Using VLITP simply for creating new possibilities, new markets, or enabling existing alternatives to be reachable can also trigger much needed change. The implementation of a new technology like RFID implies a direct relationship between business and IT—something that has become of increased importance in the last decade. Airlines are a vital part of the service industry, focusing on the transportation of people, their luggage, and goods from one point to another. RFID brought into the airline industry a system that tracks the location of passengers' luggage, directly impacting the level of service an airline can provide its customer. RFID introduced new possibilities in luggage handling that are beginning to impact the entire airline industry. In the commercial airline industry, where fiercely competition has been well established, customer satisfaction and service level are important selection factors for passengers. Like its predecessor—the barcode system—RFID tracks luggage and is used to identify which baggage belongs to which customer but using a different technique to do so. RFID, being a lot more accurate then the barcode system, makes the decision by an airline to implement it a move to establish its critical performance indicator.

INTRODUCTION

Radio frequency identification (RFID) is the current technology of demand in the airlines industry. Nearly all airlines in Western Europe and North America are either currently implementing an enterprise system or planning to do so, in an attempt to take advantage of RFID. The chapter not only explains what RFID has to offer but also what a successful FRID implementation in the airline industry involves. Airline business remains a large and growing industry that facilitates economic growth, world trade, international investment and tourism and is therefore central to globalization taking place in many industries. Its list of evidence in successfully automated business processes presents an interesting proposition for enterprise systems. As a result of the previous approaches to using IT innovations for business process management, certain IT projects are considered a minimum requirement in this industry. For example, almost every airline uses a reservation system that provides an easy way for customers to book flights. Other departments of the airline—including planning, catering, human resources, frequently use the information generated by the sales department of an airline, etc.—to make appropriate contribution to the common goal of providing quality services to its customers. Competition among air carriers creates the need for constant cost reduction, thus, resulting to a constant search for cutting edge technology to assists in winning the battle for more customers.

In the decision making process for a VLITP, finance always plays a big role. This is especially true for VLITPs involving RFID. It is important to see the (potential) impact of RFID on the airlines industry as a whole, with focus on the understanding / fulfillment of customer needs, the overall business plan of airlines, and the financial aspects of implementing RFID. The two cases in this chapter demonstrate the importance of identifying various characteristics of competing technologies (i.e. RFID and barcode) and comparing them in costs and benefits before deciding which one to go with. Such comparison should also take into account the costs for maintenance and updates. Where it can be proven that the potential benefits are far less than the costs of the VLITP, implementation is certain not a prudent business decision.

The goal of this chapter is therefore to assist the reader in finding answers to the below questions when considering the implementation of RFID in the airline industry:

- Which technologies may emerge from the implementation of VLITPs in the airline industry?
- What is RFID?
- What are the major issues surrounding the need to introduce RFID?

- What are the technical specifications of the RFID?
- What are the major RFID implementation issues?
- What are the collaboration issues between business and IT with regard to RFID?
- What are the benefits and limitations for the RFID?
- How does RFID focus on Enterprise Wide Management?
- How does RFID concern various tasks / functions in the whole enterprise?
- How does RFID contribute to effective integration?
- How does RFID affect the well-established competition among air carriers?

The chapter includes two separate cases from the airline industry operating in The Netherlands. One major process significantly affected by both cases was the baggage handling system. More than 60% of passengers complained in the 80's and early 90's specifically about the misappropriation of baggage. This prompted the International Air Transport Association (IATA) to investigating the use of RFID in the airline industry in 1995, with the view of improving efficiency and effectiveness of baggage handling process at all major airports. The positive and conclusive result led to the decision by many airports to use RFID.

Radio Frequency Identification

What is RFID?

Wyld et al (2005) describe RFID as a technology that uses microchip in a tag (also called transponders) or label to store data. The data are transmitted from or written to the tag or label when it is exposed to radio waves of the correct frequency and with the correct communications protocols from an RFID reader. This enables the possibility to trace the movement of the bag from check-in to the time it enters the plane (Morgenroth, 2003). RFID therefore encompasses the process of wireless identification of objects using radio frequency waves (Vinet et al, 2004).

The Need for RFID

Before RFID many airports used barcode system. Barcodes are printed on the luggage tags, along with the destination and transfer airports so that humans can read them as well (Morgenroth, 2003). The use of barcodes didn't prove to be completely satisfactory because they could only read about 90% of the bags correctly. That meant the airlines devoted considerable time, energy and money manually intervening in the process of reuniting passengers with their bags when the systems didn't work properly (Wyld et al, 2005). This also implies that barcodes require an accompanying cost in terms of personnel—which often proved inappropriate for preventing mishandled of baggage. Airlines managements were continuously searching for a better technology to eliminate the rising cost of compensation and hotel bills as a result of mishandling customers'. By mid-90s the annual costs were calculated at over $100 million (Strategic direction, 2006).

Technical Specifications of RFID

An RFID system consists of the following components:

- Tags with an antenna.
- Reader with an antenna.
- A software application.

The RFID reader sends out electromagnetic waves which the tag antenna is tuned to receive. The tag modulates the waves and sends it back to the reader, which converts the new waves into digital data. The RFID readers have wireless connections to the system containing a software application through a LAN.

Major RFID Implementation Issues

Hardgrave and Miller (2006) suggest that managers make little effort in trying to gain full insight in RFID technology before implementation. This would help their understanding of the impact of having such technology on the organization strategy:

- What would be possible?
- What could not be expected from the information system?
- Clear definition of the business objectives.
- Metrics for the success of this RFID application.

From a financial point of view, the organization should also consider elements such as direct and indirect costs, return on investment and payback time in acquiring RFID systems. There are further technical aspects that should be taken into consideration before implementing RFID systems. Some of these technical aspects are the optimum tag placement and orientation, choosing the cost efficient tags, and being cautious about system alterations are important implementation issues.

Benefits and Limitations of RFID

Airline customers expectation from a good service is not only limited to a safe flight, but also arrival at their destinations on time and being accompanied by their luggage within a reasonable time. RFID can decrease levels of lost or misplaced luggage and thereby influencing customer (perceived) satisfaction levels. The overriding benefits of RFID focus primarily on improving efficiency, accuracy, security, and cost savings that will allow air carriers to be more competitive in the long run (Kelly and Erickson, 2005). One of the major benefits of implementing RFID in airline is to improve customer service automatically by reducing the amount of complaints directing to mishandled baggage. The system also contributes benefits of matching passengers to their baggage, which also helps to reduce security threats (Morgenroth, 2003). Every airline that implements RFID enjoys guaranteed immediate benefits of a significant reduction in time required for processing baggage. Subsequently, RFID would reduce the number of mishandled baggage resulting in lower mishandling costs.

Such reduction in airline operational costs would have to be measured against the very high costs of implementing an RFID system. As a result, smaller airlines are very hesitant to directly implementing RFID. Instead, several smaller airlines have taken an alternative measure of outsourcing their baggage processing functions to large airlines that have successfully implemented RFID (i.e. passengers for Aer Lingus baggage are handled by KLM at Schiphol). Due to the growing opportunities in airline industry, it is envisage that benefits from implementing RFID would not be limited to baggage handling alone.

KLM ROYAL DUTCH AIRLINES

This section begins with a short overview of KLM history, presenting an insight into the move to be the first European air carrier to embrace RFID technology into its airline business. That will be followed by a description of the condition of operation before RFID implementation, before reviewing the implementation process and unfolding important lessons leant by KLM during this process.

KLM is an international airline company operating globally with a home based in Schiphol International Airport, near Amsterdam in The Netherlands. KLM reported full yearly revenue of €24.11 billion, including €1.4 billion operating income, with net earnings per share of €2.63 in March 2008 (Air France-KLM, 2008). This truly Dutch company was founded in 1919, making it the oldest international airline in the world. Under the previous name, Amsterdam-London air link, its inaugural flight was in 1920 when it had a yearly high quality travel service for

345 passengers and 25,000kg (55,000lbs) of mail and cargo (KLM Group, 2008) compare to 74.8 million passengers in the year ending March 2008 (Air France-KLM, 2008). KLM is still serving the oldest air route in Europe 90 years later, but occasionally using Boeing 747 capable of carrying more than 400 passengers plus 20,000kg (44,000lbs) of mail and cargo. By 1924 the first KLM flight took off for Indonesia, then appropriately named Dutch East Indies (KLM Group, 2008). This truly adventurous journey took weeks to reach Indonesia compared to the 15 hours journey from Amsterdam to Indonesia in the 21st Century. Growth has always been part of KLM strategy objective. In 1946, KLM was the first European airline after World War II to launch an air service across the Atlantic Ocean, from Amsterdam to New York. The next innovation came in 1960 when KLM took delivery of its first jet aircraft. It was not only larger than the earlier types of planes but also flew much faster (KLM Group, 2008). As a result of this much higher speed, travelling times became much shorter. This in turn made the world seem much smaller, because it became possible to reach almost any international airport in the world within 24 hours.

Structure of the Company

KLM now forms the core of the KLM Group, which includes other brands like City-hopper and Transavia. In addition to being part of a group of airlines companies, KLM and Air France joint in mid-2004 to form the world's largest airline group—according to 2003 revenues figures quoted in Riseley (2004) with the official company name "Air France-KLM Group". Figure 14.1 below shows a schematic overview of the new combination of business groups:

The scope of our research that underpins this chapter was limited to KLM only and does not include the entire KLM Group. KLM contains three distinguished departments at the core of its business operations: KLM Passenger Services, KLM Engineering and Maintenance, and KLM Cargo. They support the four core activities underpinning KLM existence: passenger transport, cargo transport, engineering

Figure 14.1. Overview of top-level structure for Air France—KLM Group

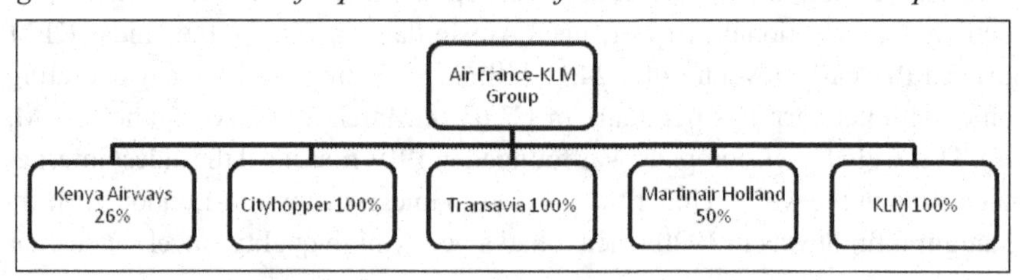

and maintenance, and charter/low cost flights (Viaene and Cumps, 2005: 542). For the purpose of this chapter and the most main department of KLM dramatically affected by the implementation of RFID is Passenger Services.

In the 2003-04 financial year KLM set out to reduce its overall internal cost base structurally. The goal was to enable the company to provide better services to its customers at lower cost and make up for the then declining operation margins in the airline industry. KLM intended to implement this cost reduction through a combination of process change, productivity gains and product improvements. KLM predicted that cost reduction alone would not guarantee more profitability (Viaene and Cumps, 2005) neither would it ensure the long-term health and growth of its business (Riseley, 2004). The management of KLM therefore decide to set a new goal where strategic orientation would be directed more towards differentiating itself from its competitors by focusing on a more direct relationship with its customers (Viaene and Cumps, 2005). This strategic goal was interpreted into delivering a more personalized and consistent customer experience at every point of contact. KLM then started to invest and implement customer relationship management (CRM) as a way to differentiate itself from its competitors.

Enterprise Systems

In the 1990s several innovations in IT led to the development of a range of software applications aimed at integrating the flow of information throughout a company. These commercial software packages were known as Enterprise Systems. During this period one particular type of enterprise systems, called ERP, caught the attention of some of the world's largest companies. Enterprise Resource Planning (ERP) systems are comprehensive packaged software solutions, which aim for a total integration of all business processes and functions (Parr and Shanks, 2000, p. 1). Parr and Shanks (2000) suggest the major advantage of these systems is the provision of a common integrated software platform for business processes. By integrating all business processes an organization can improve its efficiency and effectiveness in operations. It has been estimated that businesses around the world have been spending almost $10 billion per year on enterprise systems (O'Brien, 1999). ERP aims to integrate business processes through the support of an integrated computer information system.

Shanks et al (2003) describe enterprise systems as large-scale organizational systems that are built around packaged enterprise system software (ESS). ESS is a set of packaged applications software modules, with an integrated architecture that can be used by organizations as their primary engine for integrating data, processes and information technology, in real time across internal and external value chains. ESS includes: Enterprise resource planning (ERP), Customer relationship management (CRM), Supply chain management (SCM), Product life cycle management (PLM), Enterprise application integration (EAI), Data warehousing and decision support, Intelligent presentation layer and eProcurement / eMarketplace / electronic exchange software.

Taking advantage of the number of possibilities offered by the Internet—including easy and fast exchange of information between divisions—enterprise systems contributes to the world becomes smaller as companies get better connected both internally and externally. Enterprise systems facilitate the sharing of specific information in only a few seconds. They put information systems in place that can yield more information more quickly that they have ever known. Enterprise systems allow managers to monitor key figures of a business in real time and they do not have to wait monthly or quarterly reports. According to Davenport (2000: p.2) enterprise systems possess the following characteristics:

- Large-scale organizational systems (consisting of people, process and IT)
- Integrated, enterprise-wide, packaged software
- Impound deep knowledge of business practices
- Semi-finished product that must be customized
- Business and IT mangers must work together.

Enterprise systems also promote globalisation and internationalisation of business operations. This usually happens when organization use enterprise system for business integration and for managing supply chain.

Possible Benefits

Companies use many different ways to reduce costs with each saving helping towards the competitive nature of the business. Davenport (2000: 69) presents these saving options via using enterprise systems:

- Savings from new approaches to work
- Savings from dismantling legacy systems
- Revenue enhancement benefits.

The following benefits can be directly realised from the use of enterprise systems:

- **Operational benefits:** Different processes can be adjusted to each other, i.e. the maintenance of planes can be scheduled more efficiently when flight information is always available.

- **Managerial benefits:** Easy to see company results (i.e. number of passengers) in real time
- **Strategic benefits:** Easy to implement pricing based on demand, the more demand for a specific flight, the higher the price.
- **IT infrastructure benefits:** It is not needed anymore to connect different IT systems from different departments because less system are used. Legacy systems can be dismantled.
- **Organizational benefits:** Less employees needed for ticketing and booking flights.

The use of ES does not always result in lower system cost, but companies can also achieve benefits from the business perspective. When processes can be executed more efficiently—like the case of booking flight with KLM—VLITP results in higher customer satisfaction and subsequently to increase in the number of customers.

Limitations and Risks

Of course, besides all the above benefits there are certain unavoidable risks companies exposed themselves to by implementing enterprise systems. Some of such risks are:

- Ending up late
- Project running over budget
- Different strategic approaches for technical and business departments.

Yusuf (2004) suggested enterprise systems implementation issues to be grouped into three areas as follows:

- **Cultural problems:** The whole company has to change their way of work, using the new system requiring everyone to integrate within their individual line of work.
- **Business problems:** The structure in the company may sometime need to change as well. This sometimes causes a very rigid company to face problems in achieving its business goals.
- **Technical problems:** The old data does not always fit in the new system.

Enterprise System at KLM

One of the major issues surrounding the need to introduce RFID was change in legislation introduced in December 1992. The European Union passed legislation to deregulate the airline industry. This meant that strong competition came into the airline industry because every European carrier could fly from any place to another and could demand landing spots. Also price fighters such as Ryan-Air and Easy-Jet were opening new segments, attracting new costumers and taking market share from KLM (Viaene and Cumps, 2005).

These resulted to passengers leaving KLM and KLM having an excess capacity. KLM then stared to struggle to meets its relatively high fixed cost of loans and leases. The response to that business condition was a need to drastically cut costs. But reducing cost wasn't going to save KLM and wouldn't give any guarantee for profit in the long run. A better solution was for KLM to differentiate by offering a more superior customer experience. This transition was to come about through the implementation of an enterprise system that would affect the complete Circle of Contracts (Viaene and Cumps, 2005).

KLM's first experience with an enterprise system was in 1997. This project failed because there was no support for the project from the business managers. It was considered to be a technical (IT) project that would be the full prerogative of the IT department. KLM's next attempt at implementing an enterprise system came as after the World Trade Center attack in New York—September 11, 2001. Thereafter KLM needed to improve the security, which would bring high investments and rising cost, to regain the confidence of customers in the airline industry. In such a climate, KLM had to reduce the cost again.

In short KLM suffered great competition and had very high costs. In the mist of these changes, KLM saw an opportunity to differentiate and bring customers a superior product. To realize this KLM had to introduce enterprise system. Another major issue surrounding the need to introduce enterprise system was a change of the marketplace. The following market factors largely affected the decision to introduce enterprise system (Riseley, 2004):

- **Increased buyer power:** Customers were becoming increasingly knowledgeable about travel products, and had the tools to evaluate the market more easily.
- **Deregulation and growing competition:** Opening European skies to any airline based in the EU had started to bring a new wave of low cost competitors. Extending deregulation could result in more direct competition from larger distance.
- **Technology:** A range of new technologies and applications made it possible to interact with large numbers of customers in far more personalized ways.
- Disintermediation: Using the Internet, airlines could sell to and support an increasing number of customers directly.

- **Commoditization:** Alliances, code-sharing, growing competition from low cost airlines, and increasingly standardized products meant that for many travellers, on short-haul routes in particular, airline seat was simply a commodity.

KLM vendor selection process started 2002 with a short list of 12 serve providers, from which Epiphany was selected to provide the management software and set up a customer database. The database would be able to support general querying and reporting on costumer data; campaign set-up and executing; and enhanced e-mailing. That was the result of Epiphany's demonstration of total commitment to realise KLM enterprise system's goals as well as guarantees for post implementation support. The new enterprise system integrated all databases—including sales data; data on costumer complaints; Operations databases; corporate costumer data; individual costumer data and so on (Viaene and Cumps, 2005). KLM chose IBM WebSphere technology for the B2B Internet portal with the enterprise system KLM could achieve the following: (Riseley, 2004):

- Identify customers' value segments
- Understand customers needs and preferences
- Create targeted marketing and sales campaigns for specific customers
- Monitor customer responses
- Apply experience to future campaigns
- Steer customers' buying and travelling behaviour

In 2007 KLM introduced TIBCO software to serve as the 'enterprise backbone' to its corporate system. This would integrate all parts of KLM systems in System Oriented Architecture platform in an effort to improve the efficiency of KLM business functionalities.

The most essential processes in KLM organization relating to passenger transfer are as follows:

- Buy a ticket (economy or business class) → Register name.
- Checking in (desks or self-service) → Check name and assign seat in airplane, label baggage with code.
- Boarding → Check tickets.
- Baggage → Transfer to destination by conveyer.
- Transfer → Passengers from other carriers that continue their flight with KLM. Baggage must not get lost in the process.

KLM makes a flight schedule—a scheme by when passengers are flown to specific locations at a particular time. The information required includes: time, gate, size of the plane, and all the decisions about the practical materials to be carried along with the passengers. As a very large organisation a single process like this one going wrong could cost the company a lot of money. Consequently it is important that processes are executed in the right way. Even little delays in the process of servicing its passenger (i.e. getting luggage to a flight few minutes late or losing one suitcase) could result in high additional cost to KLM. The need to meet complete accuracy in the airline servicing process makes it compelling to have good enterprise system. Because various aspects of this process involve risks, the process has to be registered and reviewed at every stage. A few examples of risk register are as follows:

- The possible failure or inability to align goals through conflicting directions within the organization.
- The non-delivery or non-availability of reliable IT hardware and infrastructure both before and during implementation.
- The resistance of change to new process methods by management and supervision.
- Management and supervision may treat the project as merely an IT implementation, rather than change in process methods.
- The project may impact on company interim and end of year accounts.
- Possible failure to cut over to the new system through an inability to load data.
- Possible failure to cut over to the new system through the inappropriate systems testing of volume, stress and data conversion.
- Inadequately educating the workforce to operate the new system properly.

Technical Specifications

Enterprise systems provide KLM with a customer-centric business strategy with the goal of maximizing profitability, revenue and customer satisfaction. Technologies that support this business purpose include the capture, storage and analysis of customer, vendor, partner and internal process information. Technology to support CRM initiatives must be integrated as part of an overall customer-centric strategy. Many CRM initiatives have failed because implementation was limited to software installation without alignment to a customer-centric strategy.

The technology requirements of a CRM strategy must be guided by an overall view of characteristics of customers and what value they expect from engaging with the organization. The basic building blocks can be implemented at different time but eventually need to be coordinated dynamically, are:

- A database for customer lifecycle information about each customer and prospect and their interactions with the organization.
- Customer intelligence: Translating customer needs and profitability projection into software that tracks whether the desired plan is followed or not, and whether the desired outcomes are obtained.
- Business modeling, customer relationship strategy, goals and outcomes: Numbers, models and descriptions of whether goals were met.
- Learning and competency management systems: Training and improving processes and technology that enable the organization to get closer to achieving the desired results.
- Analytics and quality monitoring: To determine profitability of customer relationship policies and activities over the lifecycle of each group of customers sharing a defined set of characteristics.
- Collaboration and social networks: Profiling and interactive technology that allows the customers to interact with the business and their fellow customers and others: prospective customers, strategic partners.

KLM currently uses SAP/R3 package, a client/server system, which consists of three tiers: The client interface, the application layer and the database layer. KLM implemented the following SAP modules (KLMSS, 2008):

- FI (Financial Accounting)
- CO (Controlling)
- MM (Materials Management)
- SD (Sales and Distribution)
- PP (Production Planning)
- PS (Project System)

Viaene and Cumps (2005) describe KLM's enterprise systems vision and mission into three pillars:

First Pillar: Projects focused on developing skills and capabilities for enhancing the operational side of customer service delivery with more personalization and consistency.

- Redesign and simplify the call center process
- Redesign complaint management

- Identify customers in all interaction points
- Deliver integrated real time view of the customer in all interaction points

Second Pillar: Projects focused on developing skills and capabilities that would enable KLM to steer service delivery and decision making effectively on the basis of customer profitability.

- Optimize customer segmentation
- Optimize campaign management
- Create a customer value-based pricing model
- Get insight in purchase drivers of the customer

Third Pillar: Projects focused on facilitating and managing the change effort as KLM would move progressively towards a customer-centric organization.

- Evangelize customer-centric culture
- Set up a customer-centric performance management process

Evidence of Business Focus

Finding evidence that an enterprise system focuses on KLM business requires finding answers to these questions:

- What is the exact business in which the company is involved?
- How does IT help to make this happen?
- What level of IT integration is sufficient for the company to succeed industry?

There are elements that are visible to the end users. There are also a lot of things that happen without it being visible directly:

- The FI module in SAP handles all the accounting
- Through materials management, needed materials are automatically purchased from the suppliers.

By introducing an enterprise wide management system during the implementation of its new online booking tool, KLM eliminated the need to use a lot of different systems simultaneously—this was achieved through data integration in a single engine. A pilot of this system was held in France in February 2003. Because of its success, it is has now been rolled out in 61 different countries, with more planned for the future.

KLM also implemented an online check-in system as part of its enterprise system. This solved a long-term problem: people who travel a lot by air were complaining about long check-in queues and busy airports. What the customers needed was an easy and fast way to get around their check-ins: the online check-in system delivered such solution (Accenture, 2008). It allows a customer to print his own ticket when checking-in online. Combined with new, high-speed luggage drop off points at the airport, KLM effectively solved the problem of the long queues and thus eliminating one issue contributing to customer dissatisfaction.

At the same time, this system helps KLM reached its business goals by increasing on-line tickets sales.

- Customers can book their tickets online
- Customers can check in online
- Staff training happens online

Major Implementation Issues

Nearly every VLITP goes through several obstacles before completion. This became clear in 1997 when KLM set up its first major enterprise system spearheaded by the ICT department. The project plan included calculations consisting of limited set of rough numbers to see how much return they would get on the ICT infrastructure investment for the business. KLM management found the plan to be lacking substantial numeric value, mainly due to high forecast for technology costs. Also the business argument for that VLITP didn't support the main purpose for that initiative. As a result that particular VLITP didn't take off. In 2002 they tried to launch another VLITP this time ensuring the business focus was foremost and the target was that the objectives of this enterprise system became an important factor of the company management strategy.

While implementing the renewed VLITP they encountered several obstacles. One of these obstacles was the fact that KLM was very product and operations driven organization, which complicate a system like enterprise system, because it is difficult to change existing technical and organizational structures. This problem was overcome when the IT and business managers started to work together in some kind of harmony. KLM had learned very valuable lessons from past IT projects and decided that business should lead the VLITP, instead of the IT department, thus, using IT department as the key enabler. Another issue was that senior managers wanted to cut the budget for the CRM project. The top management decided that this project was too important to cut back on, even in the period that business was declining.

Some problems also occurred in certain areas of KLM's global operations when running CRM applications, mainly due to the lack of sufficient network bandwidth in those areas. Upgrading certain parts of the network, and also using file compression solved this. Another implementation problem KLM came across was the fact that the IT infrastructure was very complex. It was difficult to implement certain required updates or patches for standard software components. Riseley (2004) records the installation of a new version of IBM's WebSphere worldwide that required 160 separate upgrades. IT experts of KLM and the implementing company working together to solve this issue. Below is a list of ten pitfalls that were considered the most common mistakes made when implementing a enterprise systems in large organizations:

- Do not try to design the system to meet everyone's total wish list.
- Make the sales team take a prime role in the design of the system.
- When switching from the old to new system, do not expect everybody to easily pick it up.
- Do not forget to plan for staff training for all staff on the new system.
- Motivate your personnel to use the new system
- Get help to implement the system
- Make someone responsible to own the data and make sure its correct and complete
- Keep the technology as simple as you can
- Make sure the business units incorporates this system into the organization's culture to use the system
- Always seek a compromise between price and functionality versus ease of use

Collaboration Between Business and IT

Due to previously failed projects spearheaded by the IT department, senior executives decided the new CRM system had to be guided and run with the involvement of senior managers from business. The system was built on four major pillars: strategy, processes, culture and ICT (Riseley, 2004). Each pillar had its own program team member who was responsible and overall the CRM team remained accountable to the commercial division.

When the new enterprise systems started KLM initially made simple and low profile changes. By doing this they could measure the financial return more easily. A project of business and IT collaboration is a project where senior managers call selected high-value customers personally. These managers then delivered the feedback to the commercial organization. This encouraged the staff to prioritize customer-related issues. Another example of a business and IT project is the use of a customer scorecard "billboard" at the entrance of the marketing floor together with weekly reports in the internal newspaper to reinforce the importance of CRM in the company's strategy. On this gigantic board are numbers posted, like percentage of revenue coming from known customers, or number of new frequent flyer program members. These numbers encourage company-wide recognition of the results of the CRM system.

KLM decided, as a result of this system wants to build a strong relationship with their customers. By implementing this system, KLM gets a better insight in the wants and needs of their customers. Riseley (2004) believe to give staff the insight to make every customer interaction an opportunity for delivering better service and improving the traveller's experience. The main advantage is that they can adjust their supply of goods and services to the demand of the customers—key factor is e-marketing. By gathering great amounts of e-mail addresses and the attachment of interests to those addresses, the doors opened for personalized marketing and value-based segmentation. This was the result of customers increasingly managing their own transactions.

Another great benefit of implementing this system was the eradication of all the different separate databases in the enterprise. Few of the major ones integrated into the new enterprise systems were an Accounting database, a Flying Dutchman Frequent Flyer Program database and an Operations database, which all contained different kinds of useful customer information. One of the major outcomes of this VLITP was to get all those separate databases into one corporate and individual customer database. All these information became available to all sections of the enterprise (see Figure 14.1).

By going through the big database, they found that flier mileage was not the best indicator for customer value. Based on that information they could identify new indicators such as frequency of flying.

One major restriction to the system is that customer must be willing to provide very personal and detailed information to KLM. If business people made use of the Frequent Flyer Program, the marketing managers could get the wrong impression of what the customer needs were. Another limitation is privacy. How much does a company need to know about its customers? Another side effect is the volume of data to be processed, considering some of the information could also be obtained from other sources. In today's digital age, there are so many sources from which the data could be obtained that the only problem is the difficulty encountered in sorting and assembling them into one single database.

Evidence of Focus on EWM (Enterprise Wide Management)

KLM's overall objective for implementing VLITP was to become a truly customer-centric organization. This meant that all parts of the organization could take advantage of the project's outcome. The goal for this particular VLITP implementation was to enable KLM to deliver the right offer to the right customer at the right time and via the right channel. KLM launched several small campaigns targeted at specific customer segments at specific moments in time, instead of a mass marketing campaign. It also became cleared that KLM needed to identify the customer value segments and better understand their customer needs and preferences.

The business rather than ICT was the most important player in this project, but business and ICT needed to work closely together. After implementation it was very important to monitor the customer responses, so KLM could apply its experiences to future projects with the objective of steering the customers buying and travelling behaviour. This increase KLM's profitability through higher repurchasing rates and lower relative spend on marketing.

The VLITP would also move beyond the Marketing and Commercial divisions into other parts of the KLM organization. KLM's strategy has transformed the existing processes in all customer-related areas, such as marketing, distribution and ground and in-flight services. The aim is to make every customer interaction within the whole organization an opportunity for delivering a better service. The expanded scope would bring with it the need for prioritization, management, and coordination of initiatives across an even larger part of the organization. KLM is also working with Air France to install one customer database across the combined business.

Every layer in the enterprise has begun to use the system. Marketing can adjust their campaigns on the information, the sales department knows which flights are popular and can give an indication on how many seats they have to book. And by filling in complaints forms the service department can easily react to complaints and suggestions of customers.

Effective Integration

This recognized that strategic, business process and ICT changes never took place in isolation and that proactive mechanisms were necessary to get the current organization and its people to go along with the planned change. The system is currently being enhanced to facilitate marketing-oriented functions in a much more integrated enterprise systems similar to the management and marketing functions of the business. The combining of customer analytics with the data of customer transactions is likely to yield considerable value.

In general, customer information is scattered throughout many databases, knowledge bases and paper files, but with an enterprise systems, the primary repository of customer knowledge. This kind of information is too important to leave it in a scattered form. As discussed earlier, the main target of new VLITP is to build a relationship with the most valuable customers. This is only possible when there is an integration of the data on these customers. Therefore, a central customer database is essential. Before this VLITP most of the necessary customer data was separated in several databases, such as in the Accounting database, Sales database or in Operations databases. Marketing, Sales and Services independently developed KLM's ICT systems. They created their own databases, business rules and analytics, and business processes to support their own business processes. There was no integration and all these databases needed to be integrated to get a 'single view of the customer'. A complementary project (CDB project) is launched for setting up a central customer database. The CDB project was the first major step in KLM's new ICT architecture and infrastructure. For proper maintenance, issues of ownership and accountability would need to be addressed. All customer interaction points within KLM's 'Circle of Contacts' would be linked into a single view of the customer data.

VLITP is not only based on technology but also helps to combine IT/technology, people and the processes within an organization. The goal of KLM by implementing this new system is to improve the baggage process. To be able to do this they have to understand better where there might be problems. The greatest risk in VLITP is that business goals and the technology goals are not the same. KLM combines this 3 major issues of an ES namely IT, people and process. They have a management team involved with this project (people) a new system RDIF (IT) and a process (baggage process).

In the case of KLM this new system affects also empowerment. Empowerment means the ability to manage or to monitor processes. By tracking and monitoring how a process works KLM can control the process. That comes from more tracking points, and RFID can help provide that with automated reads and a higher read rate than bar codes. So the airline knows at every moment where the baggage is. This is a cut across boundaries like departments; because KLM is now able to follow the entire baggage-process across not only departments but also across the whole enterprise.

VLITPs should also take a value-chain view of business in which functional departments coordinate their work. Further out, Air France-KLM believes that the level of baggage visibility enabled by the introduction of RFID will provide the basis for new services for their passengers. (Collins, 2008). In the long term, they would be able to tell a passenger that their luggage has definitely arrived with them at the airport and exactly when it will arrive at the carousel. This will increase de satisfaction of customers.

The company envisions a separate video monitor in the arrivals hall that informs passengers when luggage will arrive in the baggage hall for collection. That would spare the organization from having to arrange to have luggage forwarded to a passenger's final destination at the company's expense. So this system leads to cost reduction. The advantages of this new ES are clear:

- Cost reduction
- Improve efficiency of internal baggage processes
- Improve processes
- Business risks limited
- Basis of new services to the passengers

This VLITP implies that KLM is committed to changing its business processes: KLM wants to reduce the lost luggage. In the beginning Air France-KLM has applied RFID tags to checked baggage at two drop-off points at Amsterdam's Schiphol airport and at Paris-Charles de Gaulle. If this system is successful it will lead to wide-scale adoption of RFID by airlines around the world. The organization did successfully adapt to the outcome of the VLITP.

DELTA AIRLINES CASE STUDY

The second case in this chapter gives a description of another situation where RFID has been implemented an airline business but this time in an American company. Delta Airlines is a air carrier that takes proud in its superior customer service, providing air transportation for passengers and cargo alike. It uses RFID technology to help in delivering a better overall travel experience to its customers.

As the 21st century unfolded Delta Airlines found itself in the favorable position of being the third largest air carrier in the US. However, due to 09/11 the airline industry as a whole faced major financial problems. The industry (in US) declined 2.6% in 2002 and the market share of Delta Airlines decreased 13% from 2000. Fixed costs throughout the industry on the other hand increased in the immediate 2 years following 09/11. Increased security costs and inflated terrorist-risk premiums drives this fixed cost increase. Labor costs were extremely high as the airlines could only regain 43.9% of total industry revenues in 2002. Even worst Delta Airlines found it had to deal with a major problem in lag time between employee layoffs and the actual realization of these cost saving measures. Delta Airlines was unable to match its competitor's lower labor cost structure as partly because of the following issues concerning higher operational costs:

- Highly paid pilots who had refused to accept a decrease in salary
- High fuel costs was beginning to be a continuing problem in the Middle East
- Government had imposed fee per passenger to cover cost of increase security measure
- Tax rates and fees had doubled in the preceding 10 years

Airport capacity, route structures, weather, technology and rising fuel and labor costs have significantly affected airline profits. On the other hand, Delta Airlines was losing market share to other low-cost carriers, which had lower costs by flying a single type of airplane and avoided the expensive networks of airports that require calculations to connect passengers and bags with flights.

Delta Airlines also faced an addition problem when its customer service was being described as poor in quality (Beals et al, 2003). That was because of the following reasons:

- Too many its passengers reached their destinations late;
- Too many incidences of baggage being considered lost;
- Instances of reported recovery were insufficient;
- Its technologies were not as customer friendly as it should have been.

While the recurring damage to morale of the entire industry was not good for business, Delta Airlines' had a particular weakness in performance that could have been handled well with the use of certain technology (Beals et al, 2003). The performance of Delta Airlines in this area was at a rate of nine lost bags per 1000 passengers. Also the rate of mishandled baggage was reports to be around 5.17 per 1000 passengers, significantly above the industry's yearly average of 4.91. Faced with all these problems, complaints about Delta Airline's services rose to 1428 (32%) in 2004 (Wyld et al, 2005).

There was obviously a need to take control of its falling reputation and get its business back on track. Delta Airlines had no other choice but to take a different approach that would lead to better services for its customers and at a discount price. The management therefore started to think about various means to its disposal that could help the airline take an operational advantage over its competitors.

New Strategies at Delta Airlines

The first step was to try and reduce operational costs. To achieve this Delta Airlines needed to make operational improvement that would involve the following:

- Having more effective flight scheduling.
- Maximizing crew resources.
- Improving maintenance processes.
- Reducing aircraft turn-around times.

Improving efficiency partly achieved by investing in new technology for its distribution channels. Installing more self-servicing kiosks in airports reduced ticket price and offering more perks to customers for using Internet check-in options.

One major technology that supported this strategy was the implementation of RFID system. This system allowed non-contact reading and track bags more precisely than with existing bar-code systems. By using RFID, Delta Airlines successfully managed to improve its baggage handling, provide real-time baggage updates and provide better, faster and friendlier service (Beals et al, 2003).

Trial for the RFID-system was carried out at one airport that proved the capacity to provide a far superior reading accuracy than the legacy bar-code system as Table 14.2 demonstrates.

This illustrated the fact that RFID would allow the airline to proactive customer service steps on baggage problems. A customer would be informed of a misplaced luggage before the customer would come forward to complain about it. The ultimate goal of Delta Airlines in implementing this VLITP was to have a baggage tracking system that will have a zero mishandling rate (Wyld et al, 2005).

RFID was that technology that offered the potential to be of a proactive nature. Using the technology, airport personnel would quickly become aware if luggage has gone astray and may sometimes even be able to rectify the problem without the passenger ever knowing that it existed. At worst, any inconvenience can be significantly minimized.

To reach a successful implementation Delta Airlines needed also to increase the employee confidence in the company. Management clearly communicated the benefits of the systems and how it would improve the business. Also, the labor

Table 14.2. Results from Delta Airlines pilot RFID test in Jacksonville

Errors per 40,000 bags	*RFID*	*Barcode*
Worst case	1,320 (96.7 percent)	8,000 (80.0 percent)
Best case	80 (99.8 percent)	6,000 (85.0 percent)

force needed to become more educated on the economics of the industry helping them to fully understand why certain initiates had to be implemented. 40 hours per year of specific on-the-job training would be implemented to teach the people this new system.

With this new RFID system and new structural adjustments Delta Airlines would provide a better overall travel experience as perceived by the customer. Delta Airlines changed in a more service provider to the customer than rather a provider of physical seats or tickets. Delta Airlines tried to gain a competitive advantage to its competitors in the form of providing service to its customers, whereby the RFID system could help them with achieving this objective. The mission of Delta Airlines then was to be an air carrier with superior customer service that provided air transportation for passengers and cargo alike, utilizing low-cost carriers and regional jets throughout the United States and around the world (Beals et al, 2003).

CHAPTER SUMMARY

This chapter has analyzed the process of implementing RFID at KLM and Delta Airlines operations in The Netherlands. Air France-KLM Group has initiated a VLITP, based on the capacity of RFID, to reach out to its customers and continues to lead the way in Europe. Our analysis of the effect of RFID on organizations in the airline industry began with the former independent Dutch company (KLM) and covered Delta Airlines changes to its business strategy—thus, making it the first airline carrier to use RFID.

RFID encompasses the process of wireless identification of objects using radio frequency waves and needed to be introduced in order to eliminate the mishandling of baggage and reduces the personnel costs. The RFID reader sends out electromagnetic waves. The tag antenna is tuned to receive this wave. The tag modulates the waves and sends it back to the reader. Managers must, before implementing, consider what can the system do and what cannot be done by the system and should also think in terms of finance because implementing a RFID system is very costly but customer service is improved by reducing the amount of mishandled baggage. One thing they must not forget is that IT (RFID) and business is necessary to meet all of an organization's information needs.

Airlines are service providers that transport people, their luggage and goods from A to B. A system that tracks the position of the luggage impacts the total service of an organization. Therefore new possibilities in luggage handling can impact the organization as a whole. Due to the fact that RFID is more accurate compared to the barcode system and service level (especially lost baggage) is a critical performance indicator in this industry some organization have in fact changed their organizational direction. The potential benefit of implementing RFID for airlines is the decrease in lost luggage and increased customer satisfaction. Also personnel expenses can decrease because fewer employees will have to help customers that lost their suitcases or find their suitcase in Hawaii while they are vacationing in Barcelona.

Delta Airlines has some major issues and was therefore unable to match its competitors. To gain an operational advantage over its competitors Delta Airlines decided to take a different approach and demanded a premium service at a discount price. A new strategy involving the use of RFID helped it to meet its ultimate goal of having a baggage tracking system that will have a zero mishandling rate. With this new RFID system and new structural adjustments Delta Airlines was able to provide a better overall travel experience as perceived by its customers. This facilitated Delta Airlines to achieve its overall mission in being an air carrier with superior customer service that provides air transportation for passengers and cargo alike, utilizing low-cost carriers and regional jets throughout the United States and around the world.

The potential benefit of implementing RFID for airlines is the decrease in lost luggage and increased customer satisfaction. Also personnel expenses can decrease because fewer employees will have to help customers that lost their suitcases or find their suitcase in Hawaii while they are vacationing in Barcelona. The costs of implementing RFID consist mainly of the purchase of the system, the costs for maintenance and updates, and the costs of training personnel to handle the new system. Airlines that have recently invested in their barcode baggage handling systems will find the cost of implementing RFID very high. But as their old and expensive to maintained barcode system become obsolete and are hard to sell at a decent price as (large Western) airlines will rather invest in a RFID system. Therefore the costs and benefits of implementing RFID are different for each airline.

The chapter has also shown how KLM tried to implement an enterprise system twice. After failing at the first attempt, KLM learned from their mistakes and made the second implementation a success—which lead to KLM wining Gartner's European CRM Excellence Award in 2004. The lessons learned by KLM of implementing the VLITP are equally valuable to other companies planning to implement a VLITP. They can be summarised as follows:

- Think big, start small and act quickly
- Keep the communications flowing and celebrate successes
- Don't confuse VLITP with loyalty programs
- Make sure business objectives, rather than IT priorities, leads the VLITP.

By following these principles of good VLITP management, both KLM and Delta Airlines have made major investments in enterprise systems. Today, both airlines are harvesting the benefits of their investment. Both companies can be described as being in a 'Competitive Process'.

REFERENCES

Accenture (2008). *High Performance at High Altitudes: KLM Royal Dutch Airlines.* http://www.accenture.com/NR/rdonlyres/air_klm_partnership.pdf. (9 Feb 2008)

Air France-KLM (2008). Financial Year 2007-08 Report for KLM. (http://coporate.airfrance.com) (Accessed 21 June 2008).

Beals, T., Tucker, M., & Vick M. (2003). A case study of Delta Airlines. *Creative Media Services*, 12 February.

Collins, J. (2008). Air France-KLM Embarks on RFID Luggage-Tag Trial. *RFID Journal.* www.rfidjourna0l.com/article/articleview (Accessed 5 Feb 2008)

Davenport, T. (2000). Mission Critical, Realizing the Promise of Enterprise Systems. *Harvard Business School.* p. 2.

Davenport, T.(2000). Mission Critical, Realizing the Promise of Enterprise Systems. *Harvard Business School Press*, p. 69.

Emerald Group (2006). Delta Air Lines and baggage handling: sky's the limit with RFID, *Strategic Direction, 22*(6), 22-24.

Hardgrave, B., & Miller, R. (2006). The myths and realities of RFID*, International Journal of Global Logistics and Supply Chain Management*, 1(1), 1- 16.

Kelly, E. P., & Erickson, G. S. (2005). RFID tags: Commercial applications versus privacy rights. *Industrial and Management Data Systems*, p.703-713.

KLM Group. (2008). History of KLM.http://www.klm.com/trave/history_of_KLM.htm. (Accessed 15 January 2008)

KLMSS. (2008). KLMSS SAP Implementation. *KLMSS.* www.klmss.com/sections/services/content/sap.ppt (Accessed 9 Feb 2008)

Morgenroth, D. (2003). Case study: RFID in the airline industry. *Focus*, April, p.52-55.

Parr, A. N., & Shanks, G. (2000). A taxonomy of ERP implementation approaches, *Hawaii International Conference on System Sciences*, p.1-10.

Riseley, M., (2004). KLM Demonstrates the Power of Persuasion to Drive CRM Success. Case study: CAS-1004-0005.

Shanks, G., P.B. Seddon and L.P. Willcocks, (2003). *Second-Wave Enterprise Resource Planning Systems: Implementing for effectiveness.* Cambridge University Press.

Smith, A. D. (2005*).* Exploring radio frequency identification technology and its impact on business systems. *Emerald,* p.16-28.

TIBCO. (2008). Air France KLM van de grond met TIBCO SOA- en CEP-software. *TIBCO.* www.tibco.com/international/netherlands/news *(Accessed* 05 Feb 2008).

Viaene, S., & Cumps, B., (2005). CRM excellence at KLM Royal Dutch Airlines. C*ommunications of the Association for Information Systems*, *16*, 539-558.

Wyld, D. C, Jones, M. A., & Totten, J. W.(2005). Where is my suitcase? RFID and airline customer service*, Marketing Intelligence and Planning*, *23*(4), 382-394.

Y. Yusuf, A. Gunasekaran, & Mark S. A. *(2004).* Enterprise information systems project implementation: A case study of ERP in Rolls-Royce, *International Journal of Production Economics, 4*, 251-266.

Chapter XV
Case Study III:
VLITP in Public Transport—Implementing OV-Chipcard in The Netherlands

ABSTRACT

Prediction markets have proven high forecasting performance in many areas such as politics, sports, and business-related fields, compared to traditional instruments such as pools or expert opinion. This case study provides details about a VLITP that achieved the goal of which makes that possible for five companies providing different modes of public transport in The Netherlands. It details the implementation problems and presents focusing point for VLITP involving multiple companies on a project that requires them to share both costs and profits. This is partly due to the need to work together and improve business practices in the same industry.

INTRODUCTION

This chapter details the implementation process of a VLITP in the Dutch public transport system—a project that have taken nearly 3 years to fully implement commonly know as OV-Chipcard. This is a good example of a VLITP implemented to satisfy a need for enterprise information systems which directly reports customer's attitude to the organization's product and services.

Several major issues surrounding the implementation of this VLITP actually originated from the conception of this project. The chapter details how a direct relationship can be made between customer attitudes to a particular company's product and the final outcome of the VLITP. Will other chapters (see Cases I and II) have managed to establish a link between VLITP implementation technique and achieving objectives, this chapter considers the basis of a success VLITP to be its wider implications to the host organization's relationship to it customer and whether the VLITP enhances that relationship or frustrates it..

The development of public transport in most European countries is the sole responsibility of government (Hermans and Stoelinga, 2003).. As such it is usually carried out on a 'piece meal' basis, completing only a small proportion at a time. Literature shows that this policy is mainly due to the massive costs of improving systems that were mostly designed for living conditions many decades ago. The transport department encouraged the creation of a joint venture named 'Trans Link Systems' to facilitate the implementation of this VLITP. The joint venture permitted the five biggest public transport companies in the Netherlands (Connexion, GVB, HTM, NS and RET) to pull resources together for this VLITP. The major objective of the VLITP was to modernise the payment method for public transport into an electronic payment method. The next section describes the OV-chipcard system, which will be followed by a brief look at the technical specification and implementation issues of this VLITP. The chapter will also describe the collaboration between business and IT managers, as well as the benefits and limitations of this strategy of VLITP implementation.

INDUSTRY SPECIFIC REQUIREMENTS VLITP

The decision to use OV-chipcard was taken seriously in 2005 when the OV-chipcard became available to the first passengers. Rotterdam was the first city in which the OV-chipcard was tested under guidance of a public transport companies called RET. RET is the public transportation company of Rotterdam and one of the five major transport companies which developed the OV-chipcard. The project was visibly promoted by director and staff using the primary process of safety, speed and comfortable means of transport for passengers using buses, underground, railway and tram. The company also investigated the issues involving proof of access, controlled movement and security supervision. These had to be completed ensuring the maintenance of material and infrastructure. The VLITP management team consisted of an Executive Board with resources belong all supporting departments, such as Finances, Management & Development and Staff & Organisation. Under the responsibility of this Executive Board fall matters such as automation, logistical, staff administration, telephone call centre, damage regulation, company restaurant and still much more.

The OV-chipcard project was perceived to benefits almost all departments and processes within the organisation. Fort the transport operators, the system represents a better way to manage passenger flow and finances. For example, it gives insight into the need for manpower, equipment and maintenance, allowing a focus on service. Operational processes, such as clearing and settling funds among transport operators, would be performed faster and more accurately. The new system the cost of printing paper tickets reduces fare evasions and decreases congestion because only OV-chipcard holders are able to proceed onto platforms.

With the arrival of the OV-chipcard the five major public transportation companies of the Netherlands will have to collaborate with each other. It was the intention that by 2009 travellers across the entire country would be in the position to use the OV-chipcard. Those five companies together would then take joint responsibility in managing all these travellers, using the facilities this VLITP would establish.

The demand for OV-Chipcard became apparent when the transport companies began to have problems with the division of income generated by selling a single paper ticket to the public that could also be used on any public transport (i.e.: trains, trams, metros and busses). The previous system also demonstrated a lack of clarity regarding the mode of transportation used by a specific passenger. That was a consequence of facilitating one ticket to be used on every mode of transportation, with the general assumption that it would have sufficient credit to cover all journeys a passenger makes within a day.

Before the implementation of Ov-Chipcard, various members of the now Tran-Linked-System sponsored separate research in an attempt to devise a more plausible solution to the problem of distributing income from ticket sales. These annual research produced results that estimated the spread of the use of public transport by a typical member of the Dutch public relaying on the use of public transport on a daily basis. Unfortunately this method proved to be very expensive and not particularly reliable. The five biggest transport companies had to combine resources for a totally reliable solution to deal with the issue of OV-chipcard. As a result of this VLITP, it is now indisputably clear what amount a specific member of the public uses, and the unused balance reminds registered on the chipcard after the passenger have check in and out at the entrances/exits of each mode of public transport. This also solves the problem of having precise statistics of total amount used by passenger on a specific mode of public transport.

There were also few other business justifications for launching this VLITP. One was to reduce the amount of passengers that got away with not paying fares when using the public transport system. The government, on the other hand, wanted to ascertain the fairness of payment on public transport as members of the public were increasingly complaining about the increase in public transport costs. The VLITP was there meant to ensure members of the public got a fairer deal by only paying for the distance covered during a particular joining through a zone charging systems.

TECHNOLOGY

The OV-chipcard is a system of electronic payments used by customers to access public transportation—Metros, trains, trams and buses—all over The Netherlands. As of the publication date of this book the OV-Chipcard was still at an introductory stage, though most public transport facilities in all the major cities were actively using it. The chipcard is used as both a payment card and an electronic ticket. Every station will eventually have a charge-up points where members of customers can check their OV-chipcard balances, as well as purchase additional credits to travel from point-A to point-B.

The cost of an OV-chipcard can also be transferred directly from a passenger's bank account, thus avoiding the need to use charge-up point at the stations. Every station will also have check-in and check-out gates at entrance and exit points so passengers can flash OV-chipcard on top of the machine which will result to the gages being opened for one passenger at a time (see Figure 15.1). A passenger must check out after previously checking into a public transport facility. This facilitates the calculation of exact fare base on distanced travelled and the system determines what amount would be credit from the chipcard.

The following briefly describes the technical specifications of the OV-chipcard:

Figure 15.1. OV-Chipcard being shown to entrance machine by a passenger

- There are 2 different types of OV-chipcard distinguished by a rechargeable card and a prepaid card. Both types of cards have a radio frequency identification chip in them which is read when a passenger enters or exists a public transport facilitate.
- Sell equipment:
    - Mobile ticket printers: used by bus drivers to print paper tickets.
    - Portable ticket printers: they are used by i.e. conductor on trains
    - Standalone selling machines for OV-chipcards
    - Selling machine at selling desks
- Card readers:
    - In the vehicle
    - Gates with integrated card readers at stations
    - Standalone card readers at stations
    - Charge-up points
- Local system: All the data collected from the above devices is stored in local systems at various stations, which are all forwarded to the Central System located at each of the main transport companies.
- Central System for each Transport Company: The system at each transport company sends raw data to the central processing system and in returns, receives back the processed data.
- Central Processing System: It receives raw data from all five public transportations companies, processes that data and sends the data back. This system is also indicates the specific figures required to complete the financial transactions between all the members of Train Linked System.

Major VLITP Implementation Issues

There were several major implementation issues concerned with this VLITP but the primary one was that of security. Open the initiation of this VLITP, OV-chipcard was identified not to be as secured as the host organizations had hoped. Despite going through the hassle of ensuring Ov-chipcard was encrypted, it soon became apparent that they may be a possibility that hackers could charge their chipcards at home.

Another issue is the apparently high number of opponents to OV-chipcard. This is mainly due to the lack of passenger's privacy. At this point there are no security guidelines for storing passengers travel data in transport companies' databases. According to the organisation that supervises personal data rights in The Netherlands (*College Bescherming Persoongegevens)*, the transport companies are not allowed to use the travel data for commercial purposes, although no other reason has been given by these transport companies for collecting passenger information relating to travel attitude.

Collaboration between both business and IT managers for this VLITP was a major issue that had to be resolved before the Ov-chipcard could be successfully implemented. The new collaboration resulted into the transport operators having a system that offers a better way to manage passengers, insight for manpower, equipment and maintenance of the public transportation. The Trans Link System performs a fast and accurate operation to get clearer and settled funds. The system's flexibility had to be adjusted because individual transport companies had rather unreliable specifications and needed a new architecture to be built that facilitated collaboration with external companies (see Chapter IX). The new architecture design made it easier for Trans Link System to modify to complete this VLITP.

There was also a technical difficulty with linking the central clearing-house system to the systems of each transport operator (i.e. Connexxion, GVB, HTM, NS and RET) and financial institution. The clearing-house is where all the date created by the different operators will be received. The system then distributes payments over each transport company, manages the automatic uploading of the ov-chipcard, and register or deregisters valid or invalid cards. This VLITP involve merging the technical, political and business requirements of the participating public transportations and financial institutions. Trans Link Systems created a business model and use-case scenario's to serve as guidance. An extensive collaboration and an innovative use of the technology are very important for this ES. This helps to create a general asset where every participant to the system (i.e. the ov-chipcard) would benefit.

Probably the highest priority for the VLITP mangers was that Trans Link Systems ensure familiarization with sufficient flexibility in all public transport operators' systems and equipment, despite the many specifications that had been built. This meant an automatic choice for the Specification Documentation on Open Architecture (SDOA). The SDOA is built to ensure that other public transport supplier can easily join the process (the smaller ones in The Netherlands). In a secure and encrypted way, all devices can continuously communicate with the ov-chipcard.

Systems Benefits

One significant benefit of this VLITP for the host organization is the cycle time reduction. A customer can easily recharge the card allowing the transport companies to automatically charge the customer's bank account. This not only avoid the time it takes for customers to stand in long queue but also allows the host organizations to collect their money directly form the passenger's account.

As a result of the ov-card being contact-less, it can be recognized from a distance of ten centimetres by the card reader. This permits a larger amount of travellers to get in and out of the public transport facilitate within a short period.

The VLITP delivers faster information transaction. The system gives the public transport operators insight in the amount of manpower, the equipment and maintenance needed within a specific period.

It also demonstrates an improvement in efficiency of the public transport processes. It gives the operators a better way to manage their customer (passengers) and improves the financial position of the operators.

Systems Limitations

Systems complexity has been given as one reason for the delay in successful implementation of this VLITP. The Trans Link System's smart card is extremely complex, mainly because it involves the integration of several technology systems. It technically consists of more than 50 different alternative fares—all programmed—and more than 60 transaction rules. The system must also take into consideration that ticket gates at stations in different cities of The Netherlands have been constructed using different specifications.

A long delay in system implemented was a general conversation of national pass-time in The Netherlands (Jansson, 2003). That's because the implementation process began in 2005 and they are still busy going through trial period. Rotterdam seems to be the only city in The Netherlands where Ov-Chipcard has been fully covered with the complete machine installation.

VLITP Management Concerns

Understanding of Customer Needs

These were the essential information needed about each customer that would use the new systems before the host organization would complete the VLITP that would satisfactorily provide public transportation to the major cities of The Netherlands:

1. Traveling from place A to place B in the city.
2. Keeping the travel time as short as possible.
3. Traveling at low costs.
4. A safe travel
5. A comfortable travel

The VLITP would need to produce an enterprise information system including the ov-chipcard that would not only possess the function for a payment method but also fulfills the function with a better understanding of the customer needs. Previous surveys, interviews and manually counting passenger were only able to create a picture of the travelers' movements, but with the Ov-chipcard other data had to be registered (i.e. like traveling time, traveling distance and the quantitative number of travelers at a certain time would need to be registered automatically). Besides the advantage of saving money by automating the research it would also give an increased amount of traveler's details to the host organization. With such a dramatic increase in customer information available to the host organization, the options for research varieties do also extend. The results can be used to create a better picture of the traveling attitude of the Dutch population and gives a better understanding of the customer needs, especially the most important one—traveling from point-A to point-B in the city and keeping the travel time as short as possible (Hermans & A. Stoelinga, 2003).

Evolving Business Plans

The new enterprise information system does intend to support the evolving of business plans. Since the public city transport company is local government property, the main goal is not gaining profit but constantly improving the city transportation. The company has to focus only on the customer needs in order to adjust the business plans (Jansson, 2003). With such an improved means of gathering information it became easier to establish and adjust business plans within the host organizations.

Managing Resources

This is particularly important to transportation companies who are constantly in need of increasing available resources. As a result of this VLITP, they would be able to make a more definite prediction about their future requirement for resources. While the short-term effect might be a mixed schedule with more or less drivers in order to respond to the predicted amount of passengers, the long-term effect, however, would be a more precise prediction indicating the number of trains, trams, and buses required at a particular point in time, piece of information that could also prove valuable for decision making on the need to add to existing fleet (Feder, 1982). With the installation of gateways at every subway stations, also being linked to OV-chipcard check in and out system, the need for ticket verification by a conductor is decreased.

Managing Value Addition

With this VLITP the host organizations can easily manage additional customer level. Considering that businesses in the transportation industry create value by branding and raising the number of travelers, this VLITP could raise the number of travelers by using valuable customer information gathered through the use of OV-chipcard. By achieving better and more tailored services, it is being suggested that more members of the public shall make the logical decision of relaying on public transportation. Something that would ultimately will result in higher value of the host organizations. It is quite difficult to improve the image of the companies operating in the transportation industry but this VLITP has continuously reported huge investments. This VLITP is therefore bring a better reputation for the host organization then expected—something that labeling billboards with the company names may probably not obtain successfully.

Monitoring & Controlling of Task/Operations

This VLITP contributes to an improved financial management process. It makes it easier to determine the number of users of the public transport because it will eventually includes all means of public transport throughout the country. Users of the system have to check in and out various public transport facilities helping the monitor and control purposes. Not only is it easy to determine the number of people who use the public transport, but also the exact number (i.e. intensity of use on specific routes, average travel range, and rush hours). It is also possible to work out more precise cost-benefit analyses on specific routes. This information could be used to assign more vehicles to certain routes and change the time schedule of the public transport to achieve an efficient distribution of capacity.

Fulfilling Customer Needs

The implementation of this VLITP will not only fulfill operational and monitoring needs of all public transport organizations, but also the additional advantage of personal registration for individual customer. Based on the personal registration of the customer, in case of a personal public transport pass, along with individually recorded travel routine it could be possible to inform the customer on specific changes that may be of relevance to that specific passenger (i.e. relevant routes off the public transport or to send out tailored advertisement, for instance by email). Is there an advantage for the public transportation company embedded in the sending out tailored advertisement? They may decide to charge a premium rate to advertisers for this more efficient advertising opportunity.

Because of the check in and check out routine, customers will pay exact fares based on the distance they have traveled, which results in a cost efficiency for the customer. An option with the OV chip system is to choose a subscription which involves an automated payment every month for the distances you have traveled. This means that you never have to recharge your public transport traveling credits, which is only possible on certain recharge points which are not available trough out the whole route.

Assess Cost Implications

One major cost necessary for this VLITP is expensive server investments necessary to implement a working and effective digital system. These are cost groups which can be identified:

- Hardware costs
- Software costs
- Costs of a non functioning system

Hardware and software are both considered start up costs for this system. Without these investments the system would never become operational. These costs consist of the development of an information infrastructure and the development, or purchase of new software. A large proportion of these costs would be directly due to the need for hardware upgrades to all transport facilities in the country, including installation of hardware around various public transports pick up points. The third cost group is related to the absence of a fully functioning system. Costs related to this are reputation costs and costs related of travelers who may not be in possession of a valid ticket. Also included in this final group of costs is maintenance to the system.

Effective Integration

The five companies formed a joint venture called Trans Link Systems. In addition to serving the customer, the system acted as a clearing-house for the host organizations. The clearing-house system receives all the data created by the various operators. The system then allocates payments across each transport company, manages the automatic uploading of the smart card, and registers/deregisters valid/invalid cards. Linking the central clearing-house system to the systems of each transport operator and financial institution was a real hurdle.

Trans Link Systems' smart card operation is extremely complex and involves the integration of numerous technology systems. As a measure of the system's complexity, more than 50 different fare products had to be programmed into the system, as well as more than 60 transaction rules. The system has to accommodate for the fact that ticket gates at stations in different cities were built to different specifications. There were many unanticipated challenges for this VLITP. The VLITP managers were concerned about potential vandalism and therefore resulted to a redesign of the plastic covers of electronic readers. Developers had to alter the card readers' laser beams when, during testing, rain and snow interfered with the ability to function.

Just as important as the technical difficulties was the high-level challenge of coordinating and managing the activities of the consortium member companies that designed, built, tested, deployed and are operating the landmark system. That challenge was matched by the task of negotiating the sometimes conflicting needs of the five major public transport companies involved. However, the managing of the different costs for each supplier of public transport isn't part of the systems' scope of application.

This VLITP presents a classis example of the doing business that it cuts across organisational boundaries. The sales department uses the OV-chipcard system to administer the sales and turnover, but at the same time the logistics department is able to use the system to analyse how much trains or busses must be brought into action on a certain moment of the day.

However, the system also cuts across boundaries of different companies as a whole. The OV-chipcard registers how much a certain person uses the public transport, and which supplier he has chosen. A citizen of Rotterdam would for instance use the services of RET (who normally supplies public transport in the Rotterdam area), so the fare paid by this customer for the journey must be given to RET. However, a person who travels by train, with the same smartcard to the North East of the country would have generated fare to the national railway company called NS (Nederlandse Spoorwegen). That brings a form of complexity that would normally be difficult to solve manually. The systems however, calculate the proportion of travels to deal with the division of fares amongst public transport providers.

CHAPTER SUMMARY

This chapter has detailed the OV-chipcard system being used for automatic payment in the public transport of The Netherlands. It provides a better understanding of the meeting individual costumers needs within a VLITP, when that customer is being served by different providers as in the case of national public transport (train, tram, bus and metro). The objective is this VLITP was to deliver a system that easily registers the travelling time, the travelling distance and the amount of travellers at a certain time. With the participating pubic transport providers avoid the need to sponsor big research work because the results can be used to create a better picture of the traveling public as well as giving them a better understanding of the customer needs.

The VLITP also add value at the management of the transportation companies by creating value through brands and raising the number of travelers on the public transport network. With a better and more tailored service it is anticipated that more people will be using public transportation, leading to higher value of the host organizations.

REFERENCES

RET (2007). www.ret.nl/ov-chipkaart.nl (Assessed January 2008)

Wikipedia http://nl.wikipedia.org/wiki/OV-chipkaart. (Assessed May 2008)

http://www.translinksystems.nl/content.asp?languageID=NL&pageID=1 (Assessed April 2008)

http://www.cwhonors.org/viewCaseStudy.asp?NominationID=258 (Assessed March 2008)

Hermans, G., & Stoelinga, A. (2003). *Competition in Dutch Public Transport.* Ministry of Transport; AVV Transport Research Centre; Public Works and Water Management.

Jansson, J. O. (2003). *Key Factors For Boosting The Bus Transport Market In Medium-size towns.* Linköping University Sweden, September, Unpublished Thesis.

Feder, P.I. (1982, February). Computer Scheduling of Public Transport: Urban Passenger Vehicle and Crew Scheduling. *Technometrics*, 24(1) 87.

Glossary

Accessibility: The extent to which the user can obtain data in an appropriate format and in time for effective use; to locate data stored in a computer system or in computer-related equipment for the purpose of reading, writing, or moving data or instructions to operate the data.

Access Control Mechanism: A mechanism that limits the actions that can be performed by an authenticated person or group.

Accumulator: A special-purpose register in the central processing unit used to store the results of arithmetic operations temporarily.

Ad Management: Methodology and software that enable organizations to perform a variety of activities involved in Web advertising.

Advanced Planning And Scheduling (APS): Programs that use algorithms to identify optimal solutions to complex planning problems that are bound by constraints.

Alliances: Cooperative business arrangements between two or more businesses with complementary capabilities.

Alliance Strategy: Competitive strategy in which an organization works with business partners in partnerships, alliances, joint ventures, or virtual organization.

Analytic Style: A systematic style of perceiving information where one follows a structured, well-organized, and deductive approach in arriving at a decision.

API: Application programming interface

Application: The use of computer-based routines for specific purposes such as accounts receivable maintenance, inventory control, and new product selection. It could also be software or computer program that process data to provide output for such a purpose.

Application Control: Control designed to protect specific applications. It relates to the processing of data within the application software and includes input controls, processing controls and output controls.

Application Generator: A program that produces application software based on information submitted by the user. A software procedure produced from a description of the functions wanted by users—a type of fourth-generation language.

Application-Level Proxy: A firewall that permits requests for Web pages to move from the public Internet to the private network.

Application Programmer: Develops software, usually in third- and/or fourth generation languages, to generate reports, update records, and perform other functions involving data stored in the database.

Application Service Provider (APS): An agent or vendor who assembles the functions needed by enterprises and packages them with outsourced development, operations, maintenance, and other services; the provision of information system or computer application over the Internet that became widely practiced in the late 1990s.

Application Software: see Application.

Architecture: The structure under which an information system's hardware, software, data, and communications capabilities are put together; how a current or proposed information system operates mechanically, described by summarizing its components, the way the components are linked, and the way the components operate together. Web services architectures differ in flexibility, expandability, security, and reliability.

Artificial Intelligence: Teaching computers to accomplish tasks in a manner that is considered "intelligent," characterized by learning and making decisions; filed of research related to the demonstration of intelligence by machines, including the ability to think, see, learn, understand, and use common sense.

ASPic: An industry consortium for application service provision.

Asynchronous Communication: The sending and receiving of messages in which there is a time delay between the sending and receiving.

Attributes of Information: Characteristics of information that make the material useful to the receiver. It could be in terms of accuracy, timeliness, reliability, origin, and so forth.

Authorization: Approval to access particular files in a database or over the Web and make certain uses of the data.

Automated Clearing House (ACH): Electronic network that connects all U.S. financial institutions for the purpose of making fund transfers.

Automation: The use of machines to perform tasks that people would otherwise do.

Autonomy: A degree of discretion individuals or groups have in planning, regulating, and controlling their work.

Auxiliary Storage: Storage that supplements that main memory section of the central processing unit. Web services make use of auxiliary storage either online or off-line.

Availability of Information: The extent to which the necessary information exists in an information system and can be accessed effectively by people who need it.

B2B Portals: An information portals for business to be contacted by other businesses.

Back End: The activities that support online order-taking which usually includes fulfillment, inventory management, purchasing from suppliers, payment processing, packaging, and delivery.

Backup: The storing of one or more copies of data, in case something goes wrong; standby, substitute, or alternate components in a computer processing system that can be used in case of failure or damage to the primary component. Regular and reliable backup system is essential to Web services to protect copies of data or programs in the event of hardware failure or other emergencies.

Bandwidth: The difference between the highest and lowest frequency that can be transmitted by a telecommunications network. This is increasingly being reinterpreted as the capacity of a channel in terms of bits per second.

Batch Processing: The processing of transaction in which transactions are gathered and stored for later execution.

Benchmarking: When running alternative systems, for the purpose of deciding among alternative application packages, a test application is used for simulating the anticipated volumes of input, output, and data manipulation.

Best Practice: The best methods for solving problems, often stored in the knowledge repository of an organization.

Bias: The creating of systematic inaccuracy in data due to characteristics of the process of creating, collecting, processing, or presenting the data.

Blog: A personal Web site that is open to the public.

Bottleneck: A processing slowdown that occurs in a Web services environment, usually because operations in certain activities or operations in the environment are lagging behind.

Bottom-Up Approach: An approach to system strategy which begins by identifying basic transaction and information processing needs. That is followed by the integration of those applications at each higher level in the organization to provide information for decision makers.

Bounded Rationality: This is a common practice of making decisions in a limited amount of time, based on limited information, and with limited ability to process that information.

Brick-And-Mortar Organizations: Old-economy organizations (corporations) that perform most of their business off-line, selling physical products by means of physical agents.

Bridge: The interconnections of two networks of the same type in a Web services environment. It accepts transmissions from one and directs them to appropriate locations on the other.

British Standards (BS-7799): is a Code of Practice for Information Security Management published in 1993 (and a subsequent Part 1 in 1995) in the United Kingdom with emphasis more on the development of an IS security management framework and policy, than the technical requirements of IT projects.

Broadband: A category of coaxial cable that carries multiple analogue signals simultaneously at different frequency ranges, suitable for voice, data, and video transmission.

Broadband Topology: This is a computer network topology in which every transmitted message or set of data goes to every node, although each node recognizes only messages addressed to it. The tradition examples are Ring and Star topologies.

Bug: A flaw found in a computer program that causes it to produce incorrect or inappropriate results.

Business Architecture: Organizational plans, visions, objectives, and problems, and the information required to support them.

Business Case: A written document that is used by managers to justify funding for a specific investment and also to provide the bridge between the initial plan and its execution.

Business Continuity Plan: A comprehensive plan for how the business and information systems will operate in case a disaster strikes.

Business Environment: Looking at an organization, this refers to everything outside the organization that affects its success, including competitors, suppliers, customers, regulatory agencies, demographics, and social and economic conditions.

Business Goals: The aim to achieve results that increase profitable revenue and/or market share expansion.

Business Model: A method of doing business by which an organization can generate revenue to sustain itself.

Business Plan: A written document that identifies an organization's goals and outlines how the organization intends to achieve the goals.

Business Process: A related group of steps or activities that use people, information, and other resources to create value for internal or external customers.

Business Process Management: involves the holistic approach of all systematic attempts to control and improve the implementation of a business process. It encompasses optimization of individual activities, optimization of the streamline between the activities, process change management, and change management within organization culture—with implications to strategy, governance, organization and culture.

Business Process Outsourcing: The subcontracting of a business process management to contractors outside the organization.

Business Process Reengineering: The complete overhaul and redesign of a business process using information technology.

Business Professional: A person in a business or government organization who manages other people or performs professional work in fields such as engineering, sales, manufacturing, consulting, and accounting.

Business System Planning (BSP): An information systems planning method that uses a top-down approach to identify the data necessary to run an organization. In this system, data are classified and put into a matrix to show their relation to the processes that create and use data. That data flow information is used to produce an information architecture, data management recommendations, and priorities for applications development.

Business-To-Business (B2B): E-commerce model in which all of the participants are businesses or other organizations.

Business-To-Consumer (B2C): E-commerce model in which businesses sell to individual shoppers.

Business-To-Employees (B2E): E-commerce model in which an organization delivers services, information, or products to its individual employees.

Business Value: Gauging how successfully a Web service application is being used or what a particular application of Web services has returned on its investment, or what a Web services application contributes to the organization's objectives.

Case Manager Approach: A decision support method based on the idea of finding past cases most similar to the current situation in which a decision must be made.

Centralization: The concept of locating decision-making authority, control, or resources at a limited number of locations in an organization. When applied to management, centralization is the location of decision-making authority at a relatively high level in the organization.

Certainty Factor: In an expert system, this is a number that describes the likelihood of a rule's conclusion given that its premises are true.

Champion: Within an information system environment, this refers to an individual who makes sure the system is recognized as important by others in the organization.

Change Agent: A person responsible for introducing change to an organization, such as a CIO implementing Web services into an organization. This person must prepare users for the change, introduce the changes to them, and reinforce the new system to return the organization to stability.

Channel: Data communications in Web services environment has a highway along which data travel from one location to another, such as telephone wire, coaxial cable, fiber optics, microwave transmission, satellites, and so forth.

Charge-Back System: This is an accounting system that motivates efficient system usage by assigning to user organizations costs for information systems and related resources.

Chief Executive Officer (CEO): The head of the organization who usually takes full responsibility of making sure the various parts of the organization are held accountable by someone.

Chief Information Officer (CIO): This is the head of the information systems department, usually with special responsibility for making sure the information system plan supports the business plan and provides direction for the organization's system-related efforts.

Circuit Switching: A method of moving data in the wide-area network within a Web services environment. Such communication circuits are established before communications start, and the system has continual and exclusive use of the circuit until the end of transmission.

CIS: Corporate information services directorate.

Click-And-Mortar Organization: An organization that conducts some e-commerce activities, but does its primary business in the physical world.

Clickstream Behavior: Customer movements on the Internet and what the customers are doing there.

Client-Server Architecture: An information system architecture consisting of client devices which send requests for service and server devices which perform the requested processing.

Closed System: Occasionally, there may exist an independent system within the Web services environment that is self-contained and does not interact with the environment.

Coaxial Cable: A single wire encased in insulating material and a protective metal casing. It provides much faster data transmission than twisted pair lines, free of noise and electrical interference, and can be used over long distances, such as in underground and under water cables.

Collaborative Commerce: The use of digital technologies that enable organizations to collaboratively plan, design, develop, manage, and research products, services, and innovative e-commerce applications.

Collision: The calculation of the same location for two different records while storing or retrieving data in a computer system.

Commodity Content: Information that is widely available and generally free to access on the Web.

Communication: A good communication in the Web service environment requires four essential elements: a source, a communication channel, a destination, and a message to be communicated.

Communication Network: The interconnection of multiple locations via any of the several channels available within a Web services environment.

Compatibility: The extent to which the characteristics and features of a particular technology fit with those of other technologies relevant to the situation.

Competitive Advantage: The advantage of one product or service over another in terms of cost, features, or other characteristics.

Competitive Forces Model: Model, devised by Michael Porter, which says that five major forces of competition determine industry structure and how economic value is divided among the industry players in the industry; analysis of these forces helps organization develop their competitive strategy.

Complexity: This determines how complicated a system is, and usually based on the number of differentiated components, the number of interacting components, and the nature of interactions between components.

Computer-Aided Software Engineering (CASE): CASE tools speed the system/software development process, automate tedious tasks, enforce development standards, and capture data that describe the system.

Computer Information Systems (CIS): see Information Systems.

Confidentiality: Keeping private or sensitive information from being disclosed to unauthorized individuals, entities, or processes.

Connecting for Health: As part of his effort to improve the impression of National Health Service Information Authority amongst the NHS staff, Richard Granger changed the program name to Connecting for Health in March 2004.

Connectivity: Web services offer the ability of users to interact with elements of the system freely, to connect from the computer within the Web services environment, regardless of location, time, or component design.

Consistency: Having the relevant factor remain unchanged throughout while the units are being compared.

Consortia: E-marketplaces owned by a small group of large vendors, usually in a single industry.

Consultant: The consultant's role within the implementation of Web services in an organization is to work with users who need to develop a specific application or formulate a report from the system. That person may also provide initial training and help as needed for users developing their own applications.

Content: The text, images, sound, and video that make up a Web page.

Content Management: The process of adding, revising, and removing content from a Web site to keep content fresh, accurate, compelling, and credible.

Context Diagram: Data flow diagram verifying the scope of a system by showing the sources and destinations of data used and generated.

Contingency Planning: Allows for alternative courses of action when the primary plans that have been developed don't achieve the goals of the organization.

Contingency Theory: A theory calling for management strategy to be tailored to circumstances, especially the nature of the work and the workers, the sophistication and complexity of the tools and techniques, and the external environment of the work group and overall organization. As Web services become implemented, management strategy usually needs to change.

Control: The concept of ensuring that operations and activities are occurring in accordance with plans and guidelines.

Controlling: The controlling function of managing VLITP involves the evaluation activities that project managers must perform. It is the process of determining if the VLITP's overall goals and objectives are being met. This process also includes correcting situations in which the goals and objectives have not being met.

Conversion: The process of adapting a new Web services application to fit in with an existing information system. This is usually the beginning of using a new operating system.

Cookie: A data file that is placed on a user's hard drive by a Web server, frequently without disclosure or the user's consent, which collects information about the user's activities at a site.

Copyright: An exclusive grant from the government that allows the owner to reproduce a work, in whole or in part, and to distribute, perform, or display it to the public in any form or manner, including the Internet.

Corporate Portal: A gateway for entering a corporate Web site, enabling communication, collaboration, and access to company information.

Committee of Sponsoring Organizations of the Treadway Commission (COSO) Framework: Is the most widely accepted and used framework for internal control. It is seen as the foundation for internal control within organizations based on internal control in general.

Cost: This is whatever the internal or external customer must give up to obtain, use, and maintain the product or services of an organization's process.

Cost-Benefit Analysis: This is a technique of assessing the effect of Web services in an organization by identifying the costs and benefits of the Web services applications. Analysis of costs and benefits are associated with the introduction of most new information systems into organization in the 21st century.

Cost Leadership Strategy: This is a strategy of competing on the basis of having a lower cost than one's competitors.

Critical Incident Logging: This is a technique of assessing the effect of Web services applications. This is done by recording noteworthy events during the usage and tracking events both before and after the system is in place.

Critical Mass: This is having enough users of an information system to be able to attain the desired benefits.

Critical Success Factor (CSF): This method of planning Web services implementation determines the information needs by identifying the factors that are essential to the organization's survival. It generates a database with the details about the organization's performance on a critical success factor to be analyzed by management.

Critical Theory of IT: Argues that the real issue is not the degree of improvements in IT *per se* but the variety of possible VLITP management strategies and the volume of IT projects among which businesses today must choose.

Customer-To-Business (C2B): E-commerce model in which individuals use the Internet to sell products or services to organizations, or individuals seek sellers to bid on products or services they need.

Consumer-To-Consumer (C2C): E-commerce model in which consumers sell directly to other consumers.

Customer-To-Customer (C2C): E-commerce model in which both the buyer and the seller are individuals and which involves activities such as auctions and classified ads.

Customer Relationship Management (CRM): The way in which an organization manages its relationship with its customers keeping the customers satisfied with the impression the organization continue to make.

Customization: The creation or modification of a product or service based on a specific customer's needs.

Data: Facts, ideas, or concepts that can be collected and represented electronically in digital form. Data could be captured, communicated, and processed electronically over the Web.

Data Architecture: This is the means by which data are managed to ensure reliability and access. Within Web services environment, the network architecture forms the infrastructure on which applications are based.

Database: A generalized integrated collection of data structured to model the natural relationships in the data. This also refers to a collection of files or a set of data that can be processed by several different computer programs.

Database Administrator (DBA): An individual or group whose assignment is to manage and protect the database with maximum benefit for all users.

Database Management System (DBMS): A software system that allows access to stored data by providing an interface between users or programs and the stored data.

Data Definition: The data description language defines the specifications for the form data much take to be used in a database.

Data Description Language (DDL): The language used to describe or define all or part of a database for creation or processing.

Data Encryption Standard: The standard symmetric encryption algorithm supported the National Institute of Standards and Technology and used by U.S. government agencies until October 2, 2000.

Data Flow Diagram: A graphic or pictorial description of the movement of data in and out of a system and between processes and data stores. It provides a logical view of the system and the movement and transfiguration of data in the system.

Data Manipulation Language (DML): The language used to transfer data between the database and Web services applications.

Data Model: A sufficiently detailed description of the structure of data to help a user or programmer thinks about the data.

Data Modeling: The process of identifying the types of entities in a situation, relationships between those entities, and the relevant attributes of those entities.

Data Quality: A measure of the accuracy, objectivity, accessibility, relevance, timeliness, completeness, and other characteristics that describe useful data.

Data Redundancy: The simultaneous use and modification of two or more copies of the same data.

Data Warehouse: A single, server-based data repository that allows centralized analysis, security, and control over the data.

Decentralization: The concept of locating decision-making authority, control, or resources at the level in an organization at which events are occurring.

Decision Support System (DSS): Certain types of information systems are intended to assist managers and users who must formulate decision alternatives for situations that are not well structured. This can be considered by some to be a problem-oriented information system.

Detailed Requirement Analysis: Process of creating a user-oriented description of exactly what a proposed information system will do.

Device Media Control Language (DMCL): This is a common language used by systems programmer to specify the physical storage of data in a database system, indicating space, overflow areas, and buffering to the specifics.

Diagnosis Treatment Combination (DTC): is simply an administrative code that reflects diagnosis and total treatment for a patient in hospitals—since 2006—and mental health—beginning in 2008. The code consists of fourteen digits which comprises all combinations of diagnosis and treatments. Each individual code has a price that is formed by the Dutch government. More specific, the Ministry of Health Care renews the prices for the codes each year. The prices are set on an average level of diagnosing and treatment of the patient.

Differentiation: A descriptive term used to refer to product perceived throughout the industry as having unique features in comparison to competing items.

Differentiation Strategy: Competitive strategy in which an organization offers different products, services, or features than those offered by competitors.

Digital Certificate: The verification that the holder of a public or private key is who they claim to be.

Digital Divide: The gap within a country, or between countries, between those that have information technology, particularly access to the Internet, and those that do not.

Digital Economy: An economy that is based on digital technologies, including digital communication networks, computers, software, and other related information technologies.

Digital Signature: An identifying code that can be used to authenticate the identity of the sender of a document.

Digitizing: Translating text, images, or sound into a form that can be stored electronically which allows for easier modification as and when needed.

Directing: the process—many would relate to managing VLITP—of supervising, or leading workers to accomplish the goals of the project team. In many VLITPs, directing involves making assignments, assisting workers to carry out assignments, interpreting organizational policies, and informing workers of how well they are performing.

Disintermediation: The removal of intermediate steps or persons between the user and the information system. It usually occurs when information systems are designed to be simpler to use and end-users learn to do their own computing.

Disk Operating System (DOS): An operating system with modules stored on magnetic disk, commonly used on personal computers.

Distributed Organization: This is a type of file organization usually used when data are stored on direct address devices. The addresses for storage of records are calculated by applying a randomizing algorithm to a record.

Distributed Processing Network: A set of hardware modules used for stand-along PCs, laptops or mainframe systems that are located in different physical locations. Each module can carry out stand-alone processing but can also be interconnected to share data with other locations or with a central facility.

Download: Refers to copying a portion of a file or database from the central computer system to a desktop, laptop, or departmental computer of the user.

DoH: Department of Health

DSS Generator: An element of a decision support system that combines languages, user interfaces, reporting capabilities, graphics facilities and so on for use as needed in creating a decision support system.

DSS Tool: A limited DSS generator, specializing, for example, in generating graphics, but with the capability controlled by the DSS Generator.

Dual Recording: A type of backup system use dot protect against loss of data. The same data are recorded on two storage devices simultaneously and updates are made to both copies.

Dumping: A type of backup to protect against loss of data. The database is copied periodically, which could be daily or weekly, and a log is kept between those periods of all transactions processed against the database.

Dynamic Web Content: Content at a Web site that needs to be changed continually to keep it up to date.

Dysfunctional Behavior: Behavior that interferes with the attainment of objectives. Resistance to a new system is dysfunctional behavior, often best managed by developing a system that doesn't engender resistance.

E-Banking: Various banking activities conducted from home or the road using an Internet connection.

E-Book: A book in digital form that can be read on a computer screen or on a special device.

E-Business: A broader definition of e-commerce, which includes not just the buying and selling of goods and services, but also servicing customers, collaborating with business partners, and conducting electronic transactions within an organization.

E-Cash: The digital equivalent of paper currency and coins, which enables secure and anonymous purchase of low-priced items.

E-Commerce Application: An e-commerce program for a defined end-user activity such as e-procurement, e-auction, or ordering a product online.

E-Commerce Strategy: The formulation and execution of a vision on how a new or existing organization intends to do business electronically.

E-CRM: The use of Web browsers, the Internet, and other electronic tough points in customer relationship management.

E-Gate: An integration engine for medical database/software used at different healthcare institutions in Europe.

E-Government: An e-commerce model in which a government entity buys or provides goods, services, or information to businesses or individual citizens.

E-Learning: The online delivery of information for purposes of education, training, or knowledge management.

E-Newsletter: A collection of short, informative articles sent at regular intervals by e-mail to individuals who have an interest in the newsletter's topic.

Economic View: The need to maximize project often leads to the pursuit of VLITP as a means of achieving the shareholders wish to see the maximum possible return on their investments—particularly true of large institutional investors. Business environment demonstrates a particular view of an organization which is that organizations exist to make a profit and that all other consideration are very much subservient to this.

Effectiveness: The ability of an individual or organization to do the things that need to be done.

Electronic Data Interchange (EDI): Exchanging business transaction data between organizations using electronic communications. Data are in specified formats understood by both organizations.

Electronic Health Record (EHR): Also see EMIS or EPD.

Electronic Mail: Electronic communications that eliminate the manual preparation, storage, retrieval, and manual distribution of information.

Electronic Medical Information System (EMIS): An information system in a medical organization used for access to healthcare information for monitoring healthcare records.

Electronic Patient Dossier (EPD): is the collection of applications, which are able to exchange data using the national infrastructure that follows a certain standard for exchanging data. Each application must also be certified as xIS (e.g. Hospital Information System) before it is considered part of the EPD. The EPD consists of a big database, containing links to the patient records, which are stored locally, of all actors.

Electronic Patient Records (EPR): see EPD.

Electronic Transfer Of Prescriptions (ETP): Providing a service for rapid and safe generation and transfer of prescriptions from primary care to the pharmacy of the patient's choice.

E-Markets (EM): The use of computers and telecommunications to create direct links between multiple buyers and sellers.

Encryption: The process of scrambling a message in such a way that it is difficult, expensive, or time-consuming for an unauthorized person to unscramble it.

End User: The individual who actually uses an information system or output, often a manager or staff member rather than an IS professional.

End-User Development: The development of information systems by end users rather than by information system professionals.

Enterprise Modeling: This is a technique for summarizing an organization's current information system architecture and designing a new architecture.

Enterprise Resource Planning (ERP): An integrated process of planning and managing of all resources and their use in the entire enterprise, which includes contacts with business partners.

Enterprise Information System: provides technology platform that enables various parties within organizations to integrate and coordinate their business processes. They provide a single system that is central to the organization and ensure that information can be shared across all functional levels and management hierarchies. It facilitates the combination of technology, people and process to benefit all parts of the organization.

Enterprise System Software (ESS): A set of packaged applications software modules, with an integrated architecture that can be used by organizations as their primary engine for integrating data, processes and information technology, in real time across internal and external value chains. ESS includes: Enterprise resource planning (ERP), Customer relationship management (CRM), Supply chain management (SCM), Product life cycle management (PLM), Enterprise application integration (EAI), Data warehousing and decision support, Intelligent presentation layer and eProcurement / eMarketplace / electronic exchange software.

Enterprise Web: An open environment for managing and delivering Web applications, combining services from different vendors in a technology layer that spans rival platforms and business systems.

Enterprise-Wide Compliance Performance: Managers of VLITPs often establish a robust reporting and risk analytics strategy for business executives as well as detailed operational reports that contain actionable intelligence for IT staff. The host organization can set compliance performance tolerances and measure against metrics such as vulnerability compliance, application compliance, remediation compliance, and coverage compliance.

Enterprise-Wide Management (EWM): See Enterprise-wide Compliance Performance.

Entity: An item or area of interest about which data are stored; may be a person, place, thing, or event.

Entropy: Deterioration of a system due to alack of maintenance input.

Entry Barrier: A factor in an industry that makes entering the field so difficult or expensive that existing organizations have s significant advantage. Information systems, such as computer-based airline reservation systems, can be entry barriers.

Entry Deterrent: A tactical entry barrier; may be invoked by an incumbent firm to make a new firm reconsider the decision to enter a market.

Environment: In a systems context, the environment is anything that is not a part of the system itself. Knowledge about the environment is important because of the effect it can have on system and because interactions between the system and the environment are possible.

Ergonomics: The intense application of 'user compliance' when designing technology.

Ethics: The branch of philosophy that deals with what is considered to be right and wrong.

Event Log: A technique for assessing the impact of information systems by maintaining a list (that is a log) of significant events or occurrences related to the introduction and use of a system.

Evolvability: The capability of databases and/or systems to change over time to accommodate new demands placed on them by users—an important objective of database management.

Exception Information: A comparison of actual performance against expectations.

Exception Report: A report produced only when certain events or circumstances are above or below prescribed standards or goals.

Executive Information System (EIS): Interactive system providing flexible access to information for monitoring operating results and business conditions.

Executive Support System (ESS): A computer-based information system designed to assist top-level executives in acquiring and using information needed to run the organization. Should provide an overview of all operations, details on request, and information to help identify opportunities and warn of potential problems.

Expert System: A type of information system intended to replicate the decisions of a human expert. Such system relies on manipulation of data and use of heuristics and includes knowledge base and explanation facility.

Explanation Facility: This exists in an expert system and tells the user what line of reasoning was used to develop a decision. This helps the user to decide whether the reasoning applies to current circumstances. It could also be used to explain why the system is requesting certain information from the user.

Extensible Markup Language (XML): Standard used to improve compatibility between the disparate systems of business partners by defining the meaning of data in business documents.

External Intelligence: This is a type of information required by top-level managers that includes formal information, gossip, and opinions about activities in the environment of an organization, such as competitor and industry changes.

Externally Distributed Information: Information released by the organization (i.e. annual reports to stockholders or news of a major program in a press release) which are usually reviewed by the chief executive before release.

Extranet: A network that uses a virtual printer network to link intranets in different locations over the Internet.

Factory Situation: This is considered a situation in which current systems are essential for the smooth functioning of an organization that Web services applications being developed or implementation are not intended to change how the organization competes.

Feasibility Study: An examination of the workability of an information systems project proposal in terms of its technical, economic, and human relations factors. Only a feasible project can be incorporated into a master plan for systems development by an organization.

Feature-Driven Development (FDD): A project management approach wherein a very preliminary functional specification is produced at the beginning, with the most important features being detailed and developed in the earlier iterations of the project development. FDD allows the management team to adapt as better information becomes available.

Feedback: Data or information collected and returned to a system or process so performance can be evaluated against expected performance and goals.

Feedback Loop: A loop built into an information system to sense the effect of output on the external environment and return that information to the system as an input, where adjustments can be made to meet predetermined goals.

Field Dependence: A style of perceiving information in which the individual tends emphasize the overall picture.

Field Independence: A style of perceiving information in which the individual tends to pull pieces out of the whole for analysis.

Fifth-Generation Language: A category of computer languages that have emerged in the last decade of the 20th century. They use knowledge bases with rules and facts fed in that describe a problem and arrive at a solution using artificial intelligence to associate rules, facts, and conditions rather than receiving a sequence of instructions.

File: A collection of related records that are stored together sometimes referred to as data set. The records are organized or ordered on the basis of some common factor called a key. Records may be of fixed or varying length and can be stored on different devices and storage media.

File Management: These are mainly the functions of creation, insertion, deletion, or updating of stored files and records in files. These operations are preformed on files.

Firewall: A network node consisting of both hardware and software that isolates a private network from a public network.

Five Forces: These are forces that affect an organization's competitiveness (considered to be promoted by Michael Porter): industry competitors, bargaining power of both buyers and suppliers, substitute products and the threat of new entrants to the industry.

Formal Structure: An organizational structure established to create meaningful responsibility and authority relationships with departmental boundaries and definitions of relationships between line and staff groups usually central to it.

Fourth-Generation Language: A group of nonprocedural language in which the user specifies what is to be done rather than how it should be done.

Frame: An HTML element that divides the browser window into two or more separate windows.

Frame-Based System: An expert support system that stores knowledge in frames that permit the interrelation of knowledge and can better handle complex subjects than a rule-based system.

Front-End Tool: A CASE tool that automates the early activities in the systems development process, such as producing dataflow diagrams.

Functional-Area-Information System: An information system that serves one part of the organization.

Functional Specification: An overview of the business problem addressed by a proposed system, the way business processes will change, and the project's benefits, costs, and risks.

General Control: relates to all information systems, including the applications that run on these systems. It affects access security, change management, data center operations, and disaster recovery.

Goals: Purposes or objectives that guide the operation of any system. Operations of systems are performed and controlled in such a way as to assist in attaining specified goals.

GOSIP: Difference forms of electronic transmission protocols used for transporting data over the Internet.

GP: General practitioner providing medical services at a local community level.

Graphic User Interface (GUI): This is where icons are used for interface to represent objects, a pointing device to select operations, and graphical imagery to represent relationships.

Group Support Systems (GSS): A decision support system that also offers features to support group decision making, either in conference rooms or on a computer network.

Groupware: Software products that support collaboration, over networks, among groups of people who share a common task or goal.

Growth Strategy: Competitive strategy in which an organization attempts to increase market share, acquire more customers, or sell more products and services.

Hardware: The electrical and mechanical devices that make up a computer system usually contain the equipment that is part of a computer system.

Help Desk: Information centers where users can get answers to questions, help in troubleshooting, and information on software and techniques to more efficiently use their computer.

Heuristic Style: An intuitive style of perceiving information using trail and error and readily revising plans on the basis of new information.

HICSS: Hospital integrated clinical support systems used to provide clinical directorates and specialist services with improved access to patient and clinical data.

Hierarchical Data Model: This model of representation shows relationships among entities in the database in the style of a family tree. In this form, one piece of information may relate to another piece at any level, or to many pieces at levels below it.

HIT: A request for data from a Web page or file.

HP-UX: Hewlett-Packard Unix operating environment software

HRI: Health Records Infrastructure service to access and move health record information as required.

Human Relations Era: The era in management theory history beginning in the late 1920s and early 1930s that increased the importance attached to determining job requirements and matching them with individuals' qualifications, and monitoring training needs and progress.

Hyperlink: The links that connect data notes in hypertext and enable users to automatically move from one Web page to another by clicking on a highlighted word or icon.

Hypertext: A type of data management program that is easy to use and has powerful retrieval capabilities which allows the user to create stacks of card containing data on one entity of interest and related graphics such as icons or forms to help users. These stacks can be interlinked to make a web.

Hypertext Markup Language (HTML): A programming language that uses hypertext to establish dynamic links to other documents stored in the same or other computers on the Internet or intranets.

ICRS: integrated care record system for the UK National Health Service.

Impact Evaluation: This determines how the implementation and use of a Web service application affects the organization. It can be carried out by the identification of changes directly attributable to the application.

Implementation: This the process of putting anew system into use that also includes completing training of all direct and indirect users of the system, and the actual conversion and start of regular use of the system.

Implementation Phase: A third phase of building or acquiring information system which is the main process of making a system operational in the organization.

Independent Software Vendors (ISV): Businesses that sell computer software and usually develop and implement information systems for organizations.

Industry Perspective: One of the perspectives top-level executives must keep in mind, watching the environment in which the organization operates. It includes immediate competitors, suppliers, government, and national and international competition to the Industry.

Informal Structure: Organizational relationships not shown on the organization chart but that are influential in functioning and that can be disrupted by a new information system.

Information: Data that have been processed into a meaningful form. Information adds to a representation and tells that recipient something that was not known before. What could be information for one person may not necessarily be information for anther person. Few importance characteristics of information are that it should be timely, accurate, and complete in order for information to reduce uncertainty.

Information and Communication Technologies (ICT): A category within which all computer telecommunications related activities in an organization may fall.

Informational Web Site: A Web site that does little more than provide information about the business and its products and services.

Information Center: An information systems facility within an organization aimed at facilitating end-user computing where train staff members assist users with both hardware and software systems.

Information Management: This is where VLITP managers consider the collection, use, and dissemination of information contained in a VLITP. Each system should ensure public access to records where required and appropriate.

Information Security: Policies and practices for securing information in VLITP provide the framework to protect the host organization's IS resources and assets. Such protection ensures the integrity, appropriate confidentiality, and availability of the data for VLITPs.

Information Security Policies: Regulations for practices that provide the framework to protect an organization's computer-supported resources and assets. This protection ensures the integrity, appropriate confidentiality, and availability of the data and systems of an organization.

Information Systems (IS): A system, usually computer-based, that processes data into a form that can be used by the recipient for decision-making purposes.

Information Technology: Hardware and software that perform data processing tasks, such as capturing, transmitting, storing, retrieving, manipulating, or displaying data.

Information Technology Architecture: This consists of logical and technical components that funnel the development and evolution of a collection of related systems.

Information Technology Vision (ITV): A view of information systems in an organization focused on processing efficiency and performance reliability, rather than on the information uses.

Information Utility Structure: An infrastructure to support systems use that can develop only in an organization that has considerable investment in and experience with information technology. The structure provides a centralized, standard direction and coordination among diverse parts of the system.

Infrastructure: The data architecture and network architecture that support an organization's application portfolios.

Institutional DSS: A decision support system provided as a complete application to be used on a continuing basis to address a general problem area, such as market analysis.

Institutionalism: New institutionalism has shifted the focus of analysis to a value-critical stance. They are interested in exploring how institutions embody and structure societal values. Though the old institutionalist pursued holistic analysis of whole government systems, the new version focuses upon individual institutions of political life. New institutionalists are interested in exploring how institutions embody and structure societal value unlike focusing on a particular set of values and model of government. They explore how institutions are embedded.

Integrated Care Record System (ICRS)—in a strategy document also called 'Delivering 21st Century IT Support'. ICRS is a system of 'closely coupled' electronic care records at the heart of the NHS IS modernisation programme

Integrated Services Digital Network (ISDN): A single type of network being used today by many in place of public telephone lines. It transmits all types of information, including data, voice, and image, over an all-digital network, with standard interfaces for telephones, computers, printers, etc. It can be used for high-speed facsimile, electronic and voice mail, as well as accessing services offered on the normal computer networks.

Integrated Systems: Several systems whose internal operation is closely linked.

Integrity: The accuracy, privacy, and security of stored data in Web services environment.

Intellectual Property: Creations of the mind, such as inventions, literary and artistic works, and symbols, names, images, and designs used in commerce.

Intelligent Terminal: A computer-oriented terminal that has built-in data checking capabilities and a small memory. It contains special functions that may also be built into the terminal to perform certain checks on the data or to handle certain transactions.

Interactive Computing: Computer processing in which the user communicates directly with the system to input data and instructions and to receive output.

Interactive Web Site: A Web site that provides opportunities for the customers and the business to communicate and share information.

Interface: This is a shared boundary between two systems, referring to the point at which one system's functioning ends and another system takes over.

Interface Engine: Program in an expert system that interacts with a knowledge base to formulate decisions or recommendations for decisions.

Internal Operations Information: A type of information required by top-level managers with key indicators of how the organization or a part of the organization is performing.

Internal Web Site Development: The process of building and maintaining the Web with an organization's own staff.

Internet: A public, global communications network that provides direct connectivity to anyone over a LAN via an ISP or directly via an ISP.

Internet Applications Layer: Provides support systems for the Internet economy ranging from webpage design to security.

Internet Business Service Provider (IBSP): Modern virtual organizations that provide Internet and other business services mainly over the Internet.

Internet Infrastructure Layer: Composed of organizations that provide Internet services.

Intermediary Layer: Composed of companies that are involved in the market-making process of the Internet.

Interoperability: The ability of users to exchange information in any form without difficulty or delay, and without concern for where another party is located or where data resides.

Intranet: A corporate LAN or WAN that uses Internet technology such as Web browsers and is secured behind an organization's firewalls.

Investment Management: The management of IT investment is an integrated view that provides a good way of managing life cycle of VLITPs.

IP VPN: Virtual Private Networks within a Web service environment, this offers a flexible, cost-effective and secure alternative to expensive leased lines, and is an especially effective means of exchanging critical information for employees working remotely in branch offices, at home, or on the road. It can also enable organizations to extend secure and cost-effective connectivity to customers, suppliers and business partners, who may have a huge physical distance between them.

IS Specialist: This person has the technical expertise in computer hardware, software, data management, and communications, and a working knowledge of selected business functions and ability to interact with executives, managers, and staff members. The person could also be referred to as systems analyst, system designer, development specialist, information analyst, programmer, or analyst.

Iterative Model: is an incremental software development process that was developed in response to the weaknesses of the more traditional waterfall model. It allows the developer to take advantage of lessons learned during the development of earlier versions of the system for later versions.

Interrelated Elements of VLITP are resources, time, money, and scope. They must be managed effectively and together if a VLITP is to be a success.

IT Threat: is any indication, circumstance, or event with the potential to cause loss of, or damage to, an asset as the result of IT. It also includes the intention and capability of an adversary to undertake actions that would be detrimental to valued assets. Adversaries might include: terrorists, either domestic or international; activist or pressure groups; criminals (e.g., white-collar, cyber hackers, organized, opportunists). Sources of threats may include: insider, external, and insiders working as colluders with external sources.

Just-In-Time Planning: A management strategy based on such careful and timely ordering that materials arrive "just in time" to meet production schedules. This saves on warehousing costs but requires close coordination with suppliers.

KLM: A Dutch royal airline with headquarters in Schiphol, near Amsterdam. The KLM Group includes other brands like City-hopper and Transavia. In mid-2004, it joint with Air France to form the world's largest airline group—according to 2003 revenues figures quoted in Riseley (2004) with the official company name "Air France-KLM Group".

Knowledge: Collected information about an area of concern.

Knowledge Base: In an expert system, the knowledge base contains specific information about the area of expertise, such as facts (data) and rules that use the facts in making a decision.

Knowledge Engineer: A person that works with experts in a particular area to learn how the experts evaluate situations, the rules of thumb they use, and how they decide what actions to take. The engineer captures that knowledge in an organized way to store in a knowledge base, for use in an expert support system.

Knowledge Management (KM): The way an organization can leverage the know-how of its employees, trading partners, and outside experts of the benefit of the organization. It can also be an essential tool for success in the highly competitive world of the global economy in the 21st century.

Knowledge Portal: A single point of access software system intended to provide timely access to information and to support communities of knowledge workers.

Legacy Systems: Older systems that have become central to business operations and may be still capable of meeting these business needs; they may not require any immediate changes, or they may be in need of reengineering to meet new business needs.

Levels: The components of an information system organized hierarchically.

Leverage: Using competitive strategies that make the most of corporate strengths. This is often referred to as how an information system that makes possible better service or cost savings can be used in a strategy to give the organization a competitive advantage.

Liability: Legal responsibility for one's actions, service, or products.

Linkage: Any activity that affects the cost or effectiveness of another activity, which may be internal to an organization or with external entities.

Linkage Analysis: A top-down method of information systems planning that concentrates on the competitive advantages of information technology. Executives analyze where information systems support could better link related activities to enhance a product, service, or productivity.

Liquidity: The result of having a sufficient number of participants in the marketplace as well as sufficient transaction volume.

Load Sharing: In Web service environment, several systems units sharing the workload are more efficient than one unit, taking advantage of free time on any unit and making processors available to everyone even if one unit need repairs.

Local Area Network (LAN): A communication network that spans a single site covering a limited geographic distance and may link workstations, terminals, printers file servers or other computer equipments.

Local Service Provider (LSP): In the forth quarter of 2003, the NHSIA began to award contracts to private service providers under the National Program. LSP represented groups of IT service providers responsible for different parts of the regional divisions of the NPfIT project.

Localization: The process of converting media products developed in one country to form a culturally and linguistically acceptable in countries outside the original target market.

Logical Systems Design: The stage of systems design when functional specifications are formulated, stating what the system should do, how it should do it, and in what sequence data input, processing, out of reports, etc, should occur.

Logical View: This is often referred to as the users' view of data, focusing on data needed for applications rather than on details of storage or access.

Low-Cost Leadership: The competitive strategy of offering products or services of suitable quality at a low cost than competitors' comparable products or services.

Machine Language: A language used by the central processing unit of a computer to execute instructions and process data. Machine language instructions can be executed or processed without any translation because they are directly understandable.

Management: The act or skill of transforming resources into output to accomplish a desired result or objective.

Management Functions: Planning, organizing, staffing, controlling, and communicating activities or issues. These may involve establishing goals and the policies, procedures, and programs needed to achieve them or the measuring of performance against goals and developing procedures, to adjust goals, procedures, and activities.

Management Hierarchy: This consists of three levels: top-management, middle -management, and operating-management levels.

Management Information: This includes not only summaries of accounting information, but also textual information ranging from memos to general economic conditions to rumors and personal experience, transformed into information usable to the executive. Operating level managers need less comprehensive information, including factual details, exception reports, and accounting information.

Management Information Systems (MIS): An information system focused on supporting decisions in cases where information requirements can be identified in advance and the situation is known to recur, so that reports can be produced periodically.

Management Science Era (MSE): A time in history (that started in World War II) when management theory became important as operations research, using mathematical and statistical tools to consider complex business problems.

Management Theory: An analysis of how the complexities of business interrelate that provides a way of predicting future events, explaining past events, and understanding causes and effects.

Managers: These are people who carry out management, thus making it necessary for this book to include discussions of managers and their roles in VLITP. A manager is a person who has the authority to use his or her discretion in making decisions and the limits to this discretion indicate the manger's place on the management ladder within the organization.

Master Development Plan: A list of projects to be designed, constructed, and implemented in an organization. Each project is identified by a name and brief description of purpose which are developed by priority based on support of organization goals and objectives.

Master File: A permanent file of data pertaining to the history or current status of a factor or entity of interest to an organization. This file is periodically updated to maintain its usefulness.

Message Switching: A communications method used in a Web services environment that allows for full messages of data to be transmitted. If a message encounters a link in use, the message is temporarily stored, then forwarded when the line is free.

Middle-Management Level: Here is where you find managers that are concerned with overseeing performance in the organization and controlling activities that move the organization toward its goals. The types of issues middle-management are expected to deal with include employee training, personnel considerations, and equipment and material acquisition.

Model: An abstraction of events, in decision-making, surrounding a process, activity, or problem to remove an entity from its environment for examination without the distraction of unnecessary elements.

Model Bank: A database of models in a decision support system which are identified by a unique name and stored.

Modem: A device used to connect a computer and transmission channel that will carry data, normally used to modulate and demodulate communication signals.

Money (costs, contingencies, profit)

Monitoring: Observing for a special purpose which may involve who accesses certain records, checking the records for proper usage, and identifying people who repeatedly attempt to use the database without proper authorization is part of the process of regulating access to stored data and could be either hardware or software performance.

Monolithic: A management information system design that attempts to build a single MIS encompassing the whole organization, anticipating all needs, usually an unrealistic endeavor.

Multilist: A data structure-like list organization in which pointers connect all records with a common attribute, plus the ability to run many lists through a database in one search and to allow a single record to belong to multiple lists.

Narrative Model: A language or narrative description of the relationship among variables in a process or system.

National Health Service (NHS): A national institution in the United Kingdom (though not a single corporation) funded mainly from general taxation and national insurance contributions, responsible for delivery healthcare care and services 'free for all' at the point of delivery.

National Program for IT (NPfIT): is an initiative by the NHS, born as a result of several plans to devise a workable IS strategy. Also called National Program, this project was estimated to costs the British taxpayers £6.3 billion for additional investment in emerging technologies in the NHS over a ten years period. It was designed to connect the capabilities of modern IS to the delivery of the NHS Plan devised in 1998. The core of this strategy is to take greater control of the specification, procurement, resource management, performance management and delivery of the information and IS agenda.

Navigation: Moving through the system, from screen to screen or from page to page in a report or input form.

Negative Feedback: Data or information about system performance fed back to the system through a feedback control loop to correct performance fluctuations and help maintain the system within a critical operating range.

Network: A group of interconnected computers, workstations, or computer devices such as printers and date storage systems which can be closed together or far apart and may be linked with any data transmission channels.

Network Architecture: The infrastructure on which applications are based like the data voice transmission capabilities in an information system.

Network Data Model: A model of relationships among entities in the database like hierarchical model except that an entity can have more than one "parent" or relationship to the next higher level.

Network Management: An activity involving the monitoring of a network's internal operations and reallocating its workload to use its capacity efficiently.

NDS: A network directory services used for listing in a Novell system

NHS: National Health Service responsible for the provision of all health care and services in the United Kingdom.

NHS Information Authority (NHSIA): The NHS Information Authority previously responsible for all IT related issues in the NHS now replaced by Connecting for Health.

NICE: National Institute for Clinical Excellence

Niche: A narrow target area for a product or service within a larger market. Aiming at this market is a competitive strategy to win the business of a specific buyer group either by differentiating the product for that group or lowering costs as a result of the narrow focus.

Niche Strategy: Competitive strategy in which an organization selects a narrow-scope segment and seeks to be the best in quality, speed, or cost in that market.

Object-Oriented Programming (OOP): A style of computing programming, based on concepts of object, classes, inheritance, methods, message passing, and polymorphism, that treats data and programs as if they are tightly intertwined.

OCS: Order communication systems used for monitoring and managing activities in healthcare delivery process.

Off-Line: Equipment or devices not connected to or in direct communication with the central part of a computer system.

Offshore Outsourcing: Use of vendors in other countries, usually where labor is inexpensive, to do programming or other system development tasks.

Online: Equipment or devices connected to or directly communicating with the central part of a computer system.

Open System: Any system that interacts with its environment through input and output.

Open System Interface (OSI): The open systems interface reference model, a framework for defining network standards, regardless of technology, vendor, or country of origin.

Operating-Management Level: Managers who are essentially supervisors and form the largest group of managers in an organization. They are normally concern with schedules and deadlines, human relations, cost and quality control.

Operating System: A software that controls the operation of a computer system by providing for input/output, allocation of memory space, translation of programs, etc.

Organization For Economic Co-Operation And Development (OECD): A committee in the European Union responsible for monitoring business arrangements within the Union.

Organizational Knowledge Base: The repository for an organization's accumulated knowledge.

Organization Analysis: A technique to evaluate the impact of a recently implemented system on the organization, assessing how structure, procedure, and policies have change and how the system is used.

Organization Theory: Focuses on alternate ways to structure an organization to best utilize people and other resources, such as equipment, material, and finances as well as providing for communication of information to appropriate personnel.

Organization Transformation: The change that has taken place when an organization's business processes, structure, strategy, and procedures are completely different from the old ones.

Organizing: Organizing refers to the way the resources are allocated and VLITP resources, assigns tasks, and goes about accomplishing its goals. In the process of organizing, project managers arrange a framework that links all members of sub-projects, workers, tasks, and resources together so the organizational goals can be achieved

Organizational Culture: The aggregate attitudes in an organization concerning a certain issue.

Output: Data or information that result from processing and are made available to users. It could also be anything that is produced by a system and movement across the boundary into the environment.

Output Device: Something that receives a processing result from the processing unit and translates them into the appropriate form for user. It normally includes printer, display screens, and telephone or similar voice output devices.

Outsourcing: The subcontracting of computer system operations, telecommunications networks, and, in some cases, user support, training, and system development to contractors outside the organization.

Overflow Area: An area of storage, particularly on secondary storage devices, in which data can be stored when the main or primary storage area is already full or in use. It can sometimes be called Overflow Bucket.

Overlapped Processing: The occurrence of input, processing, and output operations simultaneously to improve throughput.

NPfIT: A national programme for information technology—a means of providing a usable electronic health record nationally to the UK.

Packet Switching: A communication method used in an Web services environment in which messages are stored in primary memory, divided into blocks, or packets, of a standard size, and transmitted. They route may be determined when the message is sent or at each node along the way.

Parallel Processing: The ability to perform multiple tasks simultaneously.

Parallel Systems: Conversion to anew information system while the old system continues operating for a period. This is usually done for the purpose of ensuring that data will not be lost if a problem arises in the new system.

Parametric User: This is a type of end-user who relies on predefined questions and structures presented by the system while entering or extracting data.

PAS: Patient administration systems

PC-OS: Personal computer operating systems

Perception: How an individual sees a situation, such as a new information system that changes information distribution with a Web services environment.

Performance Monitoring: Using monitors to determine usage of components of a system to help determine whether additional resources are needed or whether existing resources should be adjusted to improve performance.

Peripheral: This is often used to describe equipment attached to a particular network or other kinds of computer system to augment it or to make it possible to use the central processing unit. It could be an input or output device, a communications device or a secondary storage device.

Personal Computer (PC): A desktop microcomputer ranging from home computers, office desktops, and laptops that have the computing functions of large systems but are limited in speed and storage capacities.

Personal Computer Software: These are software packages designed for the personal computer commonly used for word processing, spreadsheets, data management, data communications, and graphics applications.

Personalization: The ability to tailor a product, service or Web content to specific user preferences.

Personalized Content: Web content that is prepared to match the needs and expectations of the individual visitor.

Phase-In-System: This is where the conversion of Web services (or other information systems) is done when the old system is gradually replaced by the new one in phases.

PHC/PHCT: Primary health care trust.

Physical Model: This represents the entity studied in appearance and, to some extent, in function.

Physical Systems Construction: The stage in systems development at which point the application is built.

Physical View: The way data are actually stored and organized on physical devices.

Pilot System: This is way of conversion to a Web services (or any other information system) in which only a small piece of the organization function is converted to the new system to see if problems develop.

Planning: The process of establishing goals, developing policies, procedures, and programs to achieve them.

Planning Information: A type of information required by top-level managers that describe major developments and programs being planned, including the assumptions and anticipated developments on which plans are based.

Platform: In regard to an information system, the basic type of computer and network that the system uses.

Positive Feedback: Data or information about the performance of a system that reinforces operation without change.

Privacy: This is a means of protecting computer records which guards against unauthorized distribution of data.

Privacy Protection is the subjective probability that customers of the host organization believe that their private information is fully protected according to their expected high standard.

Privacy Risk is the recognition that a potential loss is associated with releasing personal information to the service provider implementing a VLITP

Private Branch Exchange: This can be a local-area network in which the existing telephone switching system is used as the center of a star topology.

Problem Avoider: A manager who tends to knowingly reject the notion that a problem exists usually by avoiding negative information and focuses on the positive aspects of a situation. That manager tends to use planning to avoid difficult situations and impending problems.

Product Development Life Cycle: This cycle includes activities to analyze, design, build and implement a specific product or service, which often varies depending on the type of product or service being developed.

Product Manager: Within organizations with an information center, this person supervises the development, use, and support of all the applications to be shared by multiple users.

Productivity: The efficiency or output of certain task that have been specified.

Project Life Cycle (PLC): is a collection of logical stages that maps the life of a project—from beginning to end. VLITPs are broken into various stages to make the project more manageable and to reduce the risks that are involved with the project.

Project Management is a process used to accomplish organizational goals through the implementation of a project; that is, a process used to achieve what an organization wants to accomplish by undertaking a particular project.

Project Managers: Are the people to whom this management task is assigned, and it is generally thought that they achieve the desired goals through the key functions of: planning, organizing, directing and controlling.

Project Methodology: Allows businesses to maximize the value of VLITP for themselves—usually by changing focus. It is not only a mindset used by businesses to reshape their entire organizational processes, but also as a radical cultural shift for organizations.

Project Planning: Planning in a VLITP occurs in different ways and at all levels. A top-level project manager plans for different events than does a project manager who supervises a group of workers who are responsible for assembling modular homes on an assembly line. This project manager must be concerned with the overall operations of the full project, while the assembly line manager or supervisor is only responsible for the line that relates directly to that sub-project.

Project Management: A practice that helps ensure a project can be completed in a structured fashion—on time, on budget and produces the expected results.

Project Scope: The definition of what the project is supposed to accomplish and the budget (of time and money) that has been created to achieve these objectives. It is absolutely imperative that any change to the scope of the project have a matching change in budget, either time or resources.

Program Swapping: The movement of jobs between main memory and secondary storage device which happens in certain kind of system configurations where a program can be swapped in and out of memory several times before execution is completed.

Programmed Decision: A frequently recurring decision that represents a well-understood and well-structured situation, permitting the development of routines to state how the decision should be made.

Programming Language: This is a language where computer processing instructions are written or coded, in which the instructions that control the movement and processing of data are written.

Project: One of the three relational operators in a data manipulation language. Can also be a temporary endeavor undertaken to accomplish a unique purpose.

Project Goal: The result that should have occurred if the project is carried out successfully.

Projection: This is a form of resistance to a new information system in which people wrongly blame the system, rather than the training, for difficulties in using the system.

Project Management is the application of knowledge, skills, tools and techniques to project activities in order to meet project requirements.

Protocol: This is the rules that allow entities to communicate with one another, including codes, identification, and acknowledgement schemes.

Prototype: This is a working version of an information system developed to allow users to evaluate its essential features that could also be considered an experimental version of a new system.

Proxies: Special software programs that run on the gateway server and pass repackaged packets from one network to the other.

Psychology: This can be seen as aiding the cognitive 'fit' between people and the things they use. It is concern with human information processing and decision-making capabilities.

Quality of Network Service (QNS): This is where many different ways as a combination of performance, features, reliability, conformance, durability, serviceability.

Quality Economic: Involves cost models that are devised by obtaining and processing cost information for a VLITP. Such models use cost information to improve quality performance during VLITP implementation in an attempt to satisfy the host organization while reducing overall cost of the project.

Radio Frequency IDentification (RFID): Is a technology that uses microchip in a tag (also called transponders) or label to store data. The data are transmitted from or written to the tag or label when it is exposed to radio waves of the correct frequency and with the correct communications protocols from an RFID reader. This enables the possibility to trace the movement of the bag from check-in to the time it enters the plane. RFID therefore encompasses the process of wireless identification of objects using radio frequency waves.

Random Organization: A file organization for data stored on secondary storage devices like magnetic disk or magnetic drum (but not magnetic tape) where records in a file may be addresses directly without accessing any other records in the file. This can be done by determining an address for the record and then going directly to the address.

Rapid Application Development (RAD): is a software development framework that focuses on building applications in a very short amount of time. Applications can be designed and developed within 60-90 days. RAD was originally intended to describe a process of development that involves application prototyping and iterative development.

Rationality: Selecting alternative expected to yield the best results, evaluated on the basis of some system of values.

Rational Unified Process (RUP): is an iterative software development process framework created by the Rational Software Corporation, to be adaptable by the project teams for the host organization. RUP is based on a set of building blocks which describe the following: what is going to be produced, the necessary skills that are required and a step-by-step explanation describing how specific development goals are going to be achieved.

Real-Time Processing: The processing of a request in an on-line system in which the results are available soon enough to be useful in controlling or affecting the activity in which the user is involved.

Record: A group of data items that are stored together and/or used together in processing. A collection of related data items treated as a unit.

Refreeze: This is where the system analysts, or change agent, have to reinforce the new system after it has been introduced, to return the organization to stability, when introducing a new system in a Web services environment.

Regional Implementation Director (RID): A regional authority that works closely with local NHS IT professionals in overseeing the National Program implementation by LSPs.

Relational Database: This is a type of database in which the data are logically structured in relations (the tables of rows and columns represent records and data items).

Reliability: The accuracy of the picture provided by the information obtained from the system.

Remote Decision Network: A part of a GSS that brings decision making together through a network of computer rather than in a conference room, where each member has access to databases and decision support software and can see information and graphics displayed by other members.

Requirement Analysis: The stage in systems development in which systems analysts determine and describe user information needs so that design and construction can follow.

Resistance: Behavior that opposes a change, such as implementing a new information system that can be best dealt with by preventing the rise of resistance.

Resources: (people, equipment, and material);

Resource Allocation: The third stage of the model for information systems planning, consisting of developing the hard, software, data communications, facilities, personnel, and financial plans needed to execute the master development plan.

Response Time: the time that elapses between a request for data or processing and the receipt of the data or processing result.

Return on Investment (ROI): A ratio of required costs and perceived benefits of a project or an application.

Revenue Model: Description of how an organization or its e-commence project will earn revenue.

Ring Topology: A computer network structure in which each network point can communicate directly with any other point. Instead of a central computer directing communications, transmissions travel around the ring, and front-end processor at each site determines whether the message is addressed to it or should be passed along.

Risks: Foreseeable events whose occurrence could cause system degradation or failure.

Risk Assessments: consist of identifying threats and vulnerabilities to information assets and operational capabilities, ranking risk exposures, and identifying cost-effective controls.

Risk Awareness: involves promoting knowledge of security risks and educating users about security policies, procedures and responsibilities.

Risk Evaluation: involves monitoring effectiveness of controls and awareness activities through periodic evaluations.

Rule-Based System: The most common type of expert support system where the knowledge about a specific situation is represented as a set of conditions against which the facts or knowledge of a situation under evaluation can be checked.

Satisficing: Here is where a system analyst finds a course of action or strategy that is considered good enough to satisfy minimum standards of a model that reduces the complexities of trying to find the ideal solution. However, if the alternatives are difficult to find, the first acceptable one is likely to be chosen but when easy to find, minimum standards may be raised.

Scanning: A quick review of multiple external information sources, such as commercial databanks, using an executive support system.

Scenario Planning: A strategic planning methodology that generates plausible alternative futures to help decision makers identify actions that can be taken today to ensure success in the future.

Schema: A description of a database that includes a statement of the characteristics of the data and the relationship between different data elements.

Scientific Management Era: This is period in the history of management theory which began in the second century of the Industrial Revolution that aimed at maximizing productivity, and where scientific management required job standards to measure performance, encouraged management-worker cooperation, and increased managers' responsibilities.

Scope: The project size, goals, and requirements.

Secure Socket Layer (SSL): Protocol that utilizes standard certificates for authentication and data encryption to ensure privacy or confidentiality.

Security: Guarding against data destruction or tampering by controlling the rights of access to the database and ability to retrieve, change, add, or delete records.

Security Risk Management: A systematic process for determining the likelihood of various security attacks and for identifying the actions needed to prevent or mitigate those attacks.

Self-Contained System: A class or type of database management system that includes its own data description and data manipulation commands that are normally independent of any programming language.

Sequential Query Language (SQL): This is a widely used relational database language.

Serial Processing: A type of processing where one task is process at a time.

Service: The actions a seller performs for a specific customer.

Service Oriented Architecture (SOA): is a framework that offers application integration based on the concept of providing independent or loosely coupled services. It is also design philosophy describing an architecture in which the applications expose their functionalities as services. These services are remotely accessible software components that perform specific tasks.

Shareability: This situation occurs when a database resource is in use by multiple users and programs regardless of department or locations.

Shell: As regards expert support systems, shell is a development tool that includes a language for stating and managing the rules or frames that make up the knowledge base and an inference engine capable of reasoning with rule sets.

Simulation: A process of imitation of reality which is usually done for computerized experiments with proposed solutions.

Site Navigation: Aids that help visitors find the information they need quickly and easily.

SLA: Service Level Agreements for the service providers to be judged by when contracts are being assessed.

Social Engineering: A type of nontechnical attack that uses social pressures to trick computer users into compromising computer networks to which those individuals have access.

Social Responsibility: Concern of a corporation for social issues like improving the air pollution or health level in a community where they operate or their systems interact with.

Socio-Technical Era: The era that began in the 1950s in management theory history in which the goals are both high job satisfaction and technological efficiency, to avoid having technical advances limited by users who are affected adversely.

Soft Information: Personal observation, opinion, and narrative commentary.

Softkey Method: An interface between system and user with a touch-screen that allows the user to select menu choices by touching them on the screen.

Software: Computer programs that control the processing of data in a computer system. It could be user-written as well as commercially prepared programs for translating, utility routines, or a database management system.

Software Monitor: A measurement tool for testing or monitoring the operating system which could also be considered a set of executable instructions embedded in the operating system.

Sorting: Arranging data into a particular sequence to make processing easier and data less cumbersome.

SOX (Sarbanes-Oxley Act): Adopted by the US congress in the early 21st Century considered mandatory for all publicly traded companies in the US. The objective of SOX is to protect investors by improving the accuracy and reliability of corporate disclosures made pursuant to the securities laws, and for other purposes.

Specified User: An end user who queries the database for data for decision-making or accesses records to update them. The user would need to know the key words or codes to access or modify records.

Sponsor: A manager who makes sure resources are allocated for building and maintaining the system.

Spreadsheet: A computer program that replicates electronically the rows and columns of a worksheet, including arithmetic capabilities and the ability to manipulate data, widely used on personal computers.

SQL: A query language used by most popular computer database systems.

Stack: The principal unit for storing information in hypertext, a powerful database management program.

Star Topology: A computer network structure in which each node is connected directly to a central computer that determines where to send the data next.

Statistical Software: Any software application designed to perform statistical analysis.

Status Information: This is a type of information required by top-level managers to keep them abreast of current problems and crises and aware of progress in taking advantage of opportunities.

Storage: Retaining data for later use, generally records of events affecting the organization.

Storage Structure: The physical organization of data as they are stored on a physical device, pertaining to sequential, random, indexed, or list file organizations.

Strategic Business Vision: A view of information systems focused on the organization's strengths and capabilities and its opportunities, and what information technology will enable the organization to do.

Strategic Health Authorities (SHAs): Regional teams of healthcare managers/authorities, responsible for the delivery of health care and services in a particular region in the UK.

Strategic Information Systems (SIS): Information systems that play a major role in value chain of a product or service for an organization.

Strategic Planning: is long-range planning that is normally completed by top-level managers in an organization.

Strategic Situation: In assessing the significance of information systems to an organization the strategic situation is one I which information technology is critical to daily operations, and applications in development are essential to future competitive success.

Strategic Sourcing: Purchases involving long-term contracts that are usually based on private negotiations between sellers and buyers.

Strategy: A broad-based formula for how a business is going to compete, what its goals should be, and what plans and policies will be needed to carry out those goals.

Strategy Assessment: The continuous evaluation of progress toward the organization's strategic goals, resulting in corrective action and, if necessary, strategy reformulation.

Strategy Formulation: The development of strategies to exploit opportunities and manage threats in the business environment in light of corporate strengths and weaknesses.

Strategy Implementation: The development of detailed short-term plans for carrying out the projects agreed on in strategy formulation.

Strategy Initiation: The initial phase of strategic planning in which the organization examines itself and its environment.

Subsystem: A part of a larger system having all the properties of a system in its own right or one system in another system.

Supercomputer: The fastest and most expensive computer available, used for designing large equipments like supersonic jet, or used for forecasting weather.

Support Situation: In assessing the significance of information technology to an organization, the support situation is one in which information systems play an important role in support activities, although the organization could manage to function without them, and future systems may not change that situation.

Surveillance: Monitoring a situation, with executive support systems, where information is checked soon after it enters the system without waiting for a report.

Switching Costs: The expense an organization or individual incurs in lost time, expenditure of resources, and hassle, when changing from one supplier or system to another, normally used as an important factor in keeping customers whose needs are being satisfied.

SWOT Analysis: A methodology that surveys external opportunities and threats and relates them to internal strengths and weaknesses.

Synonym: This pertains to tow or more keys that derive to the same storage address under a particular key transformation algorithm.

Syntax Error: An error such as incorrect punctuation or spacing in a program that causes the program to fail that is common in third-general language.

System: An organized entity characterized by a boundary that separates it from all other system. A system might also consist of other systems or components and might interact with its environment through input and output.

System Development Environment: Every VLITP needs an environment that is good for systems development. Application of varies elements in that environment to system development can provide consistent management and control of VLITPs.

System Development Life Cycle: The activities in developing a computer systems project beginning with perception of a need for the project. This is followed by the performance of a feasibility study. Thereafter if a project is accepted, the analysis, logical and physical design, and testing stages of the project occur. After the system has been tested and all errors corrected, the implementation stage takes place. During the system use, however, the system is evaluated, possibly which usually leads to maintenance and further changes.

System Programmer: A person who handles the storing of data in a database, working with physical rather than logical view of data. The person organizes data using an agreed-on storage structure to best meet other user's needs, selects storage devices, and specifies details of storage.

System Testing: Testing a system before implementation to determine how well it will perform and whether it meets original specifications. This is usually separate and distinctly done from program testing which looks for logic errors.

Tacit Knowledge: The knowledge that is usually in the domain of subjective, cognitive, and experiential learning which is highly personal and hard to formalize.

Tactical Planning: is short-range that is done for the benefit of lower-level managers, since it is the process of developing very detailed strategies about what needs to be done, who should do it, and how it should be done

Tangible Benefits: These are benefits that can be measured directly to determine how well a system is performing.

Technician: This person maintains equipment, diagnosing malfunctions and making repairs. The person may also monitor new developments in software and evaluate them for potential use.

Telecommunications: Transmission of data from one device to another device in a different location.

Teleconferencing: The use of electronic transmission to permit two or more people to meet and discuss an idea or issue.

The Open Group Architecture Framework (TOGAF) is a detailed method and a set of supporting tools used in VLITP for developing enterprise architecture. It is published and promoted by The Open Group and may be used freely by any organization wishing to develop an enterprise architecture.

Theory Of Change: This is a scientific theory identifying three stages of any change process.

Time: Measurement of project work in terms of task durations, dependencies, and critical path.

Time Sharing: A computer system that is shared by two or more users in the same interval of time, where each use is unaffected by the others and may be unaware of anyone else's presence on the system.

Time Slice: The amount of time a particular computer program may execute in a time-sharing system. At the end of such period, control is transferred to another program and the control will alternate between programs, with each using multiple time slices in the course of a complete program execution.

Time Strategy: Competitive strategy, in which an organization treats time as a resource, then manages it and uses it as a source of competitive advantage.

Token Passing Method: This is a local-area network access method used with ring networks to avoid colliding messages. It entails a particular string of bits called the token going around the network until a device with a message to transmit picks up the token, transmits, and returns the token to the network.

Tool: Any device that improves the performance of a task. Tools for developing information systems increase developer productivity and/or enhance system quality. Tools exist for analysis, design, and development.

Top-Down Approach: An approach to systems planning that focuses on organization goals and strategies which requires a high degree of top-management involvement. It involves functions such as research and development, production, and marketing properly identified along with their information needs, and applications and databases to meet those needs that have been identified.

Topology: The arrangement of nodes in computer network (such as bus, star and ring) used for transmitting data.

Total Quality Management (TQM): This is a business strategy based on identifying, analyzing, and improving processes that directly or indirectly create value for the customer.

Tracking Changes: is the monitoring of host configuration compliance during VLITP which helps create an audit trail that may be used to demonstrate effective use of IT resources as well as levels of compliance to internal and external users.

Trainer: This is a person in the information center for any organization who directly works with users to familiarize them with multiple-user applications developed by the center, with software packages, or with high-level languages such as query and retrieval.

Transaction: This normally refers to an even that involves or affects a business or organization that takes place during the course of routine business activities.

Transaction Processing: This is where one uses information technology to increase volume, accuracy, or consistency in processing data about business transactions.

Transaction Processing System: This is the processing of data about business activities, such as sales and movement of inventory.

Transmission Control Protocol/ Internet Protocol (TCP/IP): A set of standards for sharing data between different computers running incompatible operating systems.

Trend Analysis: The study of performance over time, attempting to show future direction using forecasting methods.

Turnaround Situation: In assessing the significance of information systems to an organization, the is the point at which the current technology is important for ongoing operations, but applications being developed are essential to revitalizing the business.

Uncertainty: This is a condition under which managers lack enough information to predict outcomes of activities accurately every time. It also occurs along with increased information needs when there are more possible results of an activity, a large number of different inputs, or difficulty in achieving goals.

Unfreeze: The preparing of an organization for change by encouraging flexible attitudes usually as part of the role of a systems analyst or change agent in introducing a new information system to an organization.

User Friendly: When an information system or application is considered easy to use.

User Working Area: An area in an application program in which data are brought in by the database management system, ready for use in the form required by the application, usually residing in the computer's main memory.

Validity: An important characteristic of information which determine if the information is meaningful and relevant to the stated purpose. A invalid information,on the other hand, means it applies to a different purpose than it was collected for.

Value-Added Network: A non-dedicated communication network made available on a subscription basis where users pay only for the amount of data they transmit.

Value Chain: A set of activities relevant to the understanding of the bases of cost and potential sources of differentiation in an organization. It may consist of basic business processes like inventory, manufacturing, marketing, or services and support activities like technology development or general management.

VLITP: very large information technology project.

View: A description of selected data from a database, incorporating the relationships among data as they are used in any application program.

Virtual Private Network (VPN): A network that creates tunnels of secured data flows, using cryptography and authorization algorithms, to provide secure transport of private communications over the public Internet.

Virus: Software that can damage or destroy data or software in a computer.

Vision: A view of what can happen by using resources to transform opportunities into reality. It can also be an image that grows out of an organization's understanding of the area of interest, which could be produced by imagination and creativity.

VLITP Methodology: can be defined, as a list of activities to do that can be adapted to a particular situation, within a specific period of time. Such a list would control and lead the actions of all members of VLITP teams during the life of the project.

Voice Processing: This process allows the storage, editing, and transmission of the spoken word. It allows voice-input word processing systems and verbal interaction with information systems for production, quality control, and other materials handling.

Warning Information: This is a type of information required by top-level managers to signal that changes are occurring, either opportunities emerging or the anticipation of trouble.

Waterfall Model: is a sequential software development model (developed by W. Royce) wherein software development is seen as steadily downward flowing process (like a waterfall) through the phases of software development.

Web-Based Conferencing: Linking together people at remote locations through their computers, over the web, where they have access to their own files as well as organization databases. Within Web services environment, the individuals are allowed to have information on the screen transmitted to all others and every piece of information can be stored by any one until purged.

Web Content Management: The process of collecting, publishing, revising, and removing content from a Web site to keep content fresh, accurate, compelling, and credible.

Web Hosting Service: A dedicated Web site hosting organization that offers a wide range of hosting services and functionality to businesses of all sizes.

Web Services: An architecture enabling assembly of distributed applications from software services and tying them together.

Webcasting: A free Internet news service that broadcasts personalized news and information in categories selected by the user.

Weighted Feature Analysis: The assessment of a new information system or application which is done by weighting the system features, such as ease of use and the likelihood of error, from least to greatest importance and having users evaluate how well the system works with respect to those features.

Wide Area Network (WAN): A telecommunications network that links geographically separated locations.

Word Processor: An information system equipment that assists in preparation and communication of written, displayed or voice information, normally used to refer to input, editing, and printing or displaying of written information.

Work Group Support System: This is a type of information system used to support managers and other staff in their day-to-day activities. It may also provide electronic and voice mail or facsimile system or be used for electronic publishing.

Workstation: A computer that offers extremely powerful processing capabilities and high-quality graphics usually used in engineering design or telephone research centers.

About the Author

Matthew Waritay Guah was borned in Sanniquellie City, Nimba County, Liberia. He currently holds a British passport and lives in The Netherlands. Matthew attended St. Mary's RC School in Sanniquellie until 1983 and spent two years at Don Bosco Polytechnic in Sinkor, Monrovia. He migrated to Great Britain in 1994 and has made it a home for his family while travelling and working in different parts of the world. He was deemed to have demonstrated exemplary achievement and distinguished contributions to the business community and certificated to appear in the 2008-2009 edition of Madison Who's Who of Professionals.

Matthew Guah presently works as Assistant Professor at Erasmus School of Economics, Erasmus University Rotterdam. Research focuses in organizational issues surrounding emerging technologies (i.e. Web Services and SOA) in the healthcare and financial industries. His research interests include neo-institutional theory, socio-economic impacts of E-Business on government services delivery, resistance to IS, organizational structure, IS infrastructure, strategic planning for IS—with a more general interest in the cognitive, material and social relationships between science, technology and business as well as their implications for present-day understandings of creativity and innovation.

Matthew Guah has his PhD in Information Systems and Management Controls from Warwick Business School, MSc in Technology Management from University of Manchester, and BSc (Honours) from Salford University, in United Kingdom. He came into academia with a wealth of industrial experience spanning over ten years (including Merrill Lynch, CITI Bank, HSBC, British Airways, and United Nations). Authored books include *Managing Very Large IT Projects* and *Internet Strategy: The Road to Web Services* (with W.Currie). Recent journal publications include *JIT, ISM, IJST&M, IT&I, IJKM, IJHT&M, IJT&HI*. Editorial membership of *JCIT, SJI, JIQ, JMIS* and *IJEC*. Reviewer for major IS journals and conferences. Also a member of ERIM, AIS, UKAIS, BMiS, and BCS.

Index

O

P

Q

U

V

W